Man o' War & Secretariat

The Forbidden Comparison

Charles Justice

Open Books
PRESS

Published by Open Books Press, USA
www.openbookspress.com

An imprint of Pen & Publish, Inc.
Bloomington, Indiana
(812) 837-9226
info@PenandPublish.com
www.PenandPublish.com

Copyright © 2014 Charles Justice

All past performance data for Man o' War and Secretariat and all North American Dirt Records data reproduced herein are Copyrighted c. 2014 by Daily Racing Form, LLC and Equibase Company. Reprinted with Permission of the Copyright owner.

All rights reserved.
No part of this book may be reproduced, stored in a retrieval system, or transmitted by any means, electronic, mechanical, photocopying, recording, or otherwise, except for brief passages in connection with a review, without written permission from the author.

ISBN: 978-0-9859367-5-4

This book is printed on acid free paper.

Printed in the USA

Related books on
Thoroughbred racing
by Charles Justice

The Greatest Horse of All: A Controversy Examined
Beyond Greatness: Four Thoroughbred Legends

Dedication

In memory of my father
Charles Joseph Justice, Sr.
1914 – 1987

and my grandparents
Joseph Marion Justice
1884 – 1949
Bess Margaret Graham Justice
1884 – 1973

Tribute

The maple leaf silhouette, used as a 'text break' symbol throughout this book, is intended to represent the central emblem of the flag of Canada.

This symbol was chosen to honor two gentlemen who played indispensible roles in shaping Secretariat's amazing career – Lucien Laurin, his trainer, and Ron Turcotte, his jockey for 18 of 21 career races including the Triple Crown victories.

The maple leaf also recognizes that the Canadian International Stakes, run at Woodbine Race Track in Toronto, Ontario on October 28, 1973, was the final competition of Secretariat's career.

Contents

Foreword
General Considerations	xv
Limitations on Statistics	xv
Overview of Book	xvii
Data Sources	xix

Prologue
In the Beginning	xxiii
Gene Pools and Other Bets	xxv
The Percent of Possible Stakes Winners	xxviii
Using the Genetic Arguments for Practical Data Comparisons	xxx

Chapter One
Introduction - Establishing the Basics	1
The World of Thoroughbred Racing	3
Background Basics	4
Distance	4
Time	5
Mass versus Weight: Other Names for Impost	6
Average Speed	7
Acceleration	8
Momentum and Kinetic Energy	10

Chapter Two
The Statistician's World	11
Populations and Samples	11
Descriptive and Inferential Statistics	14

Mean or Average of a Sample	15
Range	15
Standard Deviation and Variance	15
The Normal Distribution	16
Correlation	19
Simple Linear Regression	25
Do Two Samples Differ Significantly—Student's t-Test	29
How the t-Test is used	31
Hypotheses and the t-Test	34

Chapter Three

Introduction	39
The Argument about Era Differences	39
Evidence against the Era Comparison Controversy	43
Applying the Work-Energy Theorem	45
Thoroughbred Improvements over Three Centuries	47
The Distribution of Talent among Foals	49
Track Variant as Estimator of Relative Ability within Foal Crop	50
Track Variants for Man o' War's Six-Furlong Races	52

Chapter Four

Man o' War	55
Juvenile Year Anomaly: The Sanford Stakes	60
A Closer Look: the Sanford and Four other Six-Furlong Races	62
First Anomaly of the Sophomore Year	69
Second Sophomore Anomaly: The Lawrence Realization Stakes	71
Final Sophomore—and Career—Anomaly	75
Life after Track Glory	78

Chapter Five

Secretariat	81
The Races	84
Secretariat's Sophomore Season	87
Let the Anomalies Begin	90
Big Red's Final Anomalies	97
The Arlington Invitational	97
The Whitney Stakes	99
The Marlboro Invitational	99
The Woodward	100
Man o' War Stakes	100
The Canadian International	101
Old Champions Never Die	103

Chapter Six

Juvenile Year Correlations	107
Basic Juvenile Year Data	108
Correlations with Time for Six Furlongs: Man o' War	110
Correlations with Time for Six Furlongs: Secretariat	114
Coding the Correlation Sequences	117
t-Test Results for the Ten Race Parameters	121
Six Furlongs versus all Races	125

Chapter Seven

Test for Significant Difference at Six Furlongs	129
Setting Boundaries for Man o' War's Time Adjustments	129
Unbiased Time Adjustments to the Raw Data	136
Adjusting Running Times for Data Comparability	138
The Uniform Distribution of Thoroughbred Ability: I	140
The Uniform Distribution of Thoroughbred Ability: II	142

The Sequence of Time Adjustments	144
The Adjustment Process for the Six-Furlong Races	145
Explanation of the Impost Adjust Procedure	151
Using Confidence Intervals to Refine the Impost Adjustment	154
A Second Type of Confidence Interval	157
Linear Regression and the Trapezoid Adjustment Method	159

Chapter Eight

Handicapping Man o' War and Secretariat as Juveniles	167
Freedom from Era Effect Guaranteed using z-Scores	173
Using Specific Race Results to Derive z-Scores	175
Peer Quality Factor	182
z-Scores and Time Adjustments Relative to the World Record	183

Chapter Nine

Man o' War and Secretariat—Sophomore Year Comparisons	187
Basic Sophomore Year Data	187
Correlations with Time for Nine Furlongs: Man o' War & Secretariat	189
Test for Significant Difference at Nine Furlongs	194
Shapiro-Wilk Normalcy Results: Ten Major Race Parameters	196
Adjusting Running Times for Data Comparability	198
The Sequence of Time Adjustments	200
Raw Data Comparison	202

Simulations using Raw Data	202
t-Test Results and a Reference Baseline	202
Adjustment Method 1: Comparison of Track Records	203
Simulation Results of the Adjustments	208
The Time Adjustment Giving Man o' War the Obvious Advantage	210

Chapter Ten

Adjustment Method 2: The Trapezoidal Time Adjustment	217
The Relationship of z-scores to the Nine-Furlong World Record	220
What White Stockings (or Socks) Prove	225

Chapter Eleven

An Extraordinary Triple Crown	229
Enter Two Faces of Linear Regression	230
Inter-Event Regression	230
Intra-Event Regression	232
Secretariat's Belmont Stakes: June 9, 1973	233
Signatures of Greatness	236
About Track Speed	238
A Related Issue	240
A Matter of Energy	241
Temporary Resolution of Track Speed Problem	242
Asides and Final Thoughts	243

Epilogue

Embracing the Statistical Gauntlet	247
The Challenge Continues	249
Science and Beyond	250

Coda	253
References	255

Appendices

A:	Basic Data: Man o' War and Secretariat	261
B:	Calculating the Standard Deviation and Variance	265
C:	Calculating Pearson's Product-Moment Correlation Coefficient r	267
D:	The Output Data from LINEST Linear Regression	269
E:	Reference Table: Selected t Values of a One-Tailed Distribution	275
F:	The Possible Combinations of Null and Alternate Hypotheses	281
G:	Generating the Normal Distribution with Excel	287
H:	The t-Test for Two Samples with Equal Variances	289
I:	Justification for the 50/50 Division of Track Variant	295
J:	Time versus Impost Trends: Six Standard Distances	297
K:	Calculating Confidence Intervals	299
L:	Adjusting Linear Trend to World-Record Trend	303
	Index	309

Foreword

General Considerations

This book presents a detailed comparison of the racing records of Man o' War and Secretariat. These two colts are generally considered the greatest North American Thoroughbred runners of the past century.

The term 'North American' is stressed because no attempt is made herein to consider the racing records of the many fine horses from other continents and countries. I am keenly aware that horses of equal caliber to Man o' War and Secretariat may have raced outside North America, but it presents challenge enough to concentrate on these two horses exclusively for present purposes of applying data comparison methods.

Comparing Man o' War with other horses—whether of North American or other origin—is greatly frowned upon by his fans. Indeed, a general mantra-like expression has evolved among his advocates which admonishes that horses from different eras cannot be compared. This dogmatic statement is actually one of the great motivators of my attempt herein to present a rational, systematic and objective comparison of the racing records of Man o' War and Secretariat.

Limitations on Statistics

Of necessity, basic statistics must be used to achieve an unbiased, objective and scientific comparison between or among racing records, often referred to as past performance records.

Even statistics, however, *cannot define or compare an inherent level of greatness* regarding these or any other horses. Many people wish to deny or overlook this fact.

Thus, zealous fans immediately excoriate statistics without considering that all it properly does is *allow the dispassionate comparison of data records.*

That is *all* statistics was developed over many decades to do - objectively compare data. At its worst, it is applied by statisticians who generate and use biased data. Notice that it is the human element, the statisticians, who misinterpret and misapply statistics.

The statistics as a discipline does not and cannot bias data. Statistics is a purely mathematical approach which allows pattern recognition in what otherwise might be only numeric chaos. It harbors none of the hidden agendas which may haunt the subconscious mind of the investigator applying it.

Data, as used herein, simply indicate the *results* a given horse achieved *at a given time and place and under particular conditions.* Such factual data does not, a priori, define the *absolute performance level* of which a horse was capable. The reader must always remember this.

If data transcended mere fact, then it might be reasonable to speak of comparing relative greatness. However, factual data shows only what a horse did under an isolated, specific competition.

If horse A ran one mile in 93.00 seconds while horse B ran the same mile under the same conditions in 93.45 seconds, then, obviously, horse A was declared the winner, and his winning time was, by direct subtraction, 0.45 seconds faster than horse B's.

These are the bare facts of one hypothetical situation. They state nothing about whether horse A was more capable, in general, than horse B, nor do they necessarily predict that horse

A would always beat horse B if they ran multiple races at the same distance and under identical conditions in the future.

I use and apply statistics only within the foregoing interpretive framework when comparing the records of Man o' War and Secretariat. I respect both colts highly. I favor neither. I come to this study with no ulterior motives or hidden agendas. This is a highly relevant fact.

Having stated the above, one must still realize that *multiple random factors* always interact to produce data, especially in Thoroughbred racing. Horses do not control these factors. They perform as their jockeys are capable of shaping a good ride under the pre-race instructions of the owner and trainer and under the track limitations on racing day.

The final result, barring accident, mishap or bad racing luck, depends greatly on the horses' actual health on race day. Many horses, including Secretariat, have run when they should not have been asked to run, and they lost.

Overview of Book

Following an initial overview in the Introduction, all statistical methods used herein are thoroughly explained as the text proceeds and a particular application is required. Chapters 1 thru 3 are devoted basically to these explanations.

Chapter 1 provides an elementary introduction to key terms used in Thoroughbred racing for those new to the sport. Chapters 2 and 3 explain the basic statistical concepts used throughout the book for the data comparisons. Even readers absolutely new to statistics should be enabled to read the book with understanding provided these chapters are read diligently.

Chapters 4 and 5 are biographical, being devoted to the lives and track achievements of Man o' War and Secretariat, respectively.

Chapters 6 thru 10 contain the statistical crux of the data comparisons. In these chapters the various statistical tests explained in Chapters 2 and 3 are applied.

Chapter 11 serves as an overview of what has emerged relative to the two champions, while an Epilogue provides a philosophical statement concerning the book's implications for further research into comparative Thoroughbred statistics.

The textual structure generally follows the format of my previous books on Thoroughbred racing and data comparison: *The Greatest Horse of All: A Controversy Examined* and *Beyond Greatness: Four Thoroughbred Legends*.

Hereafter when either of these books is cited, the acronyms GHA and BG are used. Similarly, DRF and ARM always mean Daily Racing Form and American Racing Manual, respectively.

It is suggested that only those potential readers particularly interested in Thoroughbred racing and the comparative study of racing records acquire this book. It is primarily for the serious student of Thoroughbred racing and data comparison. Others should not apply.

That being stated, if a reader approaches the book comfortable with eighth-grade level mathematics and with a positive, unbiased mindset, he or she will have little or no trouble understanding the basic statistical methods used herein. This is not rocket science by any stretch of the imagination!

Appendix A gives a condensed summary of the complete racing records of both Man o' War and Secretariat for easy reference, plus a table of world-record times from five furlongs thru thirteen furlongs.

As preliminary aid to those unfamiliar with Thoroughbred racing, one furlong is one-eighth of one mile. Thus, six furlongs is three-quarters of one mile, and ten furlongs is a mile and a quarter.

The statistical methods explained and used herein include: 1) applying and interpreting the *normal distribution* with detailed use of its two major parameters, the *mean* (average) and the *standard deviation*; 2) *correlation* and it interpretation; 3) hypothesis testing using the *t-test*; and 4) *linear regression* as performed by Microsoft Excel's© 2003 LINEST function.

If potential readers distinctly dislike or even fear mathematics on any level, they too are discouraged from purchasing this book. It is, to reiterate, impossible to do more than a superficial and subjective comparison of two or more Thoroughbred past performance records without at least using basic statistics.

These admonitions apply whether the horses being compared are Man o' War and Secretariat or Joe Place and Larry Show.

Enough such inadequate, qualitative comparisons—perhaps more properly called 'anecdotes'—exist and prove little, if anything. To paraphrase President Barak Obama in one of his initial campaign speeches, the world doesn't need another superficial comparison.

Data Sources

The data used herein for all analyses are taken directly from two major and trusted sources: the 2005 revised edition of *Champions* and the 2012 edition of the *American Racing Manual*. Both books are published by the Daily Racing Form.

Serious readers who wish either to check the conclusions herein or to explore additional possibilities are encouraged to obtain these books and perform their own research. Microsoft's Excel 2003 application, including Lumenaut's © add-on for the Shapiro-Wilk Normalcy check, was used exclusively to perform all analyses. Both are readily available and popular computer packages.

Readers should also consider that the data used herein are *not the author's personal possessions*. Some reviewers chose to assume as much regarding my previous books. The data herein are the *numbers recorded* as a result of two horses running fifty-three years apart—Man o' War in 1919 and 1920 and Secretariat in 1972 and 1973.

The author was not yet born when Man o' War raced, but he was fortunate to have seen Secretariat run.

As a researcher interested in comparing these data and hopefully clarifying some misunderstood issues related to these two great champions, I only used the data in standard statistical applications. I also assumed that it was accurate as I found it.

The data were in no way modified or made a means by which to advance my own favorite horse, as some might choose to conclude. I believe that an honest and close reading will obviously support that statement.

Actually, if pressed about what horse was my favorite, I might lean more toward the great Native Dancer, now largely marginalized by racing pundits. He actually had a better record than either Man o' War or Secretariat, having lost only the Kentucky Derby—and that by a head—in his 22 career races. But that is another story.

In conclusion, readers **will never find stated herein** that either Man o' War or Secretariat was the better horse. Again, as stressed in GHA and BG, data analysis can *never disclose* a horse's inherent greatness. Those who think otherwise completely misunderstand the nature of statistical methods and objectives.

Statistics only suggests the *acceptance or rejection of hypotheses*—as readers will discover if they do not already know this—concerning whether one sample seems to differ or not differ significantly from another. That is all statistics does or can do. This idea is thoroughly examined, explained and illustrated throughout the text.

If you, therefore, are someone seeking affirmation of your predilections about one of these horses versus the other, this also is not the book for you.

Such multiple caveats having been clearly stated, I wish those well who desire to enter and seriously consider the past performance records of the two Thoroughbred runners generally considered the greatest colts in the history of North American racing.

Charles Justice
Bloomington, Indiana
January 6, 2013

Each of us is all the sums he has not counted.
~ Thomas Wolfe

Prologue

Despite many gratuitous warnings that comparisons of Man o' War and Secretariat are forbidden by the very era separating them, I choose to compare their racing records herein.

I do not believe that an era difference existed on any conceivable level between Man o' War and Secretariat. I do not, in fact, believe that an era difference exists *anywhere* within Thoroughbred racing history, from the publication of the General Stud Book by James Weatherby in 1791 to the present.

This point will be expanded and illustrated as the book progresses.

In The Beginning

If one compares Man o' War's six six-furlong juvenile (two-year-old) race times to the times for Secretariat's three six-furlong juvenile races, the first and most readily gathered *fact* is that their respective averages were 72.21 s and 70.47 s, respectively, with 's' meaning seconds.

Subtracting the latter time from the former shows that 1.74 s, on average per six-furlong race, separates the two colts.

The same procedure applied to Man o' War's two nine-furlong sophomore races and Secretariat's four nine-furlong sophomore races discloses a difference between their average times of: 110.40 s—108.04 s, or 2.36 s.

Therefore, if an era difference actually existed between the colts—some ethereal quality of time's passing which rendered their races inherently different, like the proverbial apples and oranges—then it changed character by 2.36 s - 1.74 s, or 0.62 s from one comparison year to the next. It was, if you will, a very malleable era.

However, neither the colts nor their race parameters were inherently different. They were both Thoroughbred horses, descended from overlapping and intermingled ancestors (Ainslie 79) dating back to more than three centuries ago. (Mackay-Smith 173) They both ran equal *measured* distances and they were both "*timed*" for those distances by technologically similar devices.

Their running feats were recorded for posterity's admiring gaze on durable materials using suitable mineral-spirit-based substances—because they happened to consistently apply their skills better than any two North American Thoroughbreds of the past century.

The previous 333 words having been written, this book could be terminated forthwith, for my major point is established—that all the rooftop shouting, over many years by people highly perturbed that anyone would compare Man o' War to another horse, has been wasted. I continue writing mainly to develop the case.

There is not and never was an "era difference" of any substance separating Man o' War and Secretariat or any other great Thoroughbred runner. The fact that relatively small time differences, 1.74 s and 2.36 s are *all the time adjustment required* to make the average times of these champions *equivalent* belies any statement that their records cannot properly be compared.

What is meant by 'equivalent?' It means precisely, for these studies, that 1000 random simulations of both Man o' War's and Secretariat's six-furlong and nine-furlong races, using adjusted run times for Man o' War *based on the above two numbers*, predict that he would win, on average, 504 of 1000 of the six-furlong races and 484 of 1000 of the nine-furlong races—provided that Secretariat received no counterbalancing compensation.

By any reasonable standard, those results would be interpreted as judging the two colts equally adept at running the respective distances. Neither was 'better,' if a qualifying term must be used.

However, the rest of the story should be told because it's important.

Gene Pools and Other Bets

The gene pool of the Thoroughbred is subject to extensive inbreeding and has been since the Irish Hobby, the English Running Horse and the Turkoman Arabian strains were introduced and combined. (Mackay-Smith 1) Thus, its genes tend to remain relatively constant over long time periods, just ten horses having contributed over half the genes found in modern Thoroughbreds. (Cunningham 94)

Additionally, *true* evolution—if there is, in fact, such an entity—is known to take hundreds of millions of years (Meyer 33), not simply the three-hundred-plus years since the Thoroughbred was bred in earnest—and certainly not the paltry 53 years separating Man o' War and Secretariat.

These are prime reasons why it is unlikely that Thoroughbred foal crops in general differ significantly in running ability from generation to generation. The crop *size* determines that *more* gifted runners may exist in one crop than in another. However, the top levels of running ability can only go so far, regardless of the crop size. Otherwise, infinite running ability could eventually appear, and that is an absurdity.

Ever since the time of Gregor Mendel it has been known, although on less sophisticated levels, that chemical units called genes exist on material called chromosomes (colored bodies) within the cell nuclei of all mammalians. (Robinson 39)

Science thus accepts the contention that these genes are responsible for passing on traits (inheritance) from one generation to the next.

Let us then posit, for simplicity and immediate argument's sake, that *eight* such gene units are the most important for the combined traits of speed and stamina in the Thoroughbred. The existence, in other words, of these specific units is what ultimately fashioned Man o' War and Secretariat.

The number eight was chosen because it nicely mirrors a rather profound concomitant symmetry idea in physics termed the Eightfold Way. (Wikipedia 1) In so doing, it may point to a mystery beyond itself. It may also mean absolutely nothing regarding equine genetics. It is, however, sufficient to mention the fact of eight's significance to at least one scientific discipline to avoid the appearance of simply favoring numerology.

Further assume that the following specific traits, needed for speed and stamina, are those directly controlled by these eight genes: 1) *skeletal soundness* and overall flexibility; 2) *muscular strength* and tone plus rapid response ability (refractory period);

3) *heart size*, especially controlled by the now recognized X factor; 4) *lung capacity*; 5) *oxygen transport efficiency*; 6) *oxygen uptake* and utilization *efficiency;* 7) *carbon dioxide dissipation* and 8) *lactic acid dissipation* (necessary to promote extended muscular exertion.

This is not simply a fanciful list of presumed factors. Recent studies have measured several of these same parameters for Thoroughbreds running special treadmills. (Minetti, A. E., et al. 2329; Cunningham 97)

An additional and approximate, but plausible, assumption is that any given foal has the probability, due to random genetic mixing of each parent's genes at conception, of one-eighth, or 1/8, of inheriting any *one* of these eight critical genes.

For inheriting two or more of the critical genes, the **odds rise rapidly** and follow a *multiplicative rule* for having 2 through 8 genes combined in one foal: 1/8 x 1/8, or $(1/8)^2 = 1/64$; $(1/8)^3 = 1/512$; $(1/8)^4 = 1/4,096$; $(1/8)^5 = 1/32,768$; $(1/8)^6 = 1/262,144$; $(1/8)^7 = 1/2,097,152$ and $(1/8)^8 = 1/16,777,216$.

Note especially that by the time 1/4,096 is reached, the odds that any given foal in Man o' War's 1917 crop would possess this number of combined critical genes—four—have been exceeded. That is because the 1917 crop had only 1,680 foals. (ARM 719) Therefore, the crop size itself was too small, theoretically, for the chance of that many specific genes combining.

However, odds being only odds, it is *possible* that Man o' War was simply "blessed" by the genetic deal of the cards and did, in fact, have at least that many combinations of critical genes. He may even have had more. The proof of the pudding would simply be his extraordinary ability on the track.

Similarly, by the time 1/32,768 is reached (five critical genes combined), the number of foals in Secretariat's 1970 crop (24,361) has been surpassed. The natural odds should theoretically have been too great for Secretariat to have that many critical combinations. Nonetheless, as for Man o' War, Secretariat probably beat the genetic odds and may also, in fact, have had the entire complement of eight *assumed* genes. Absolutely nobody who has walked the earth knows.

The Percent of Possible Stakes Winners

The above discussion is consonant with the idea that a fixed percentage of foals within a given crop will ever have a chance of winning just a *single* stakes race.

The figures 2.5% (Hollingsworth 35) and 5% (Ours 62) are sometimes mentioned in this regard, of which details will follow in a later discussion.

Taking 2.5% of Man o' War's crop, 1,680, gives 42. The same percent of Secretariat's crop, 24,361, is 609. The number of probable (according to natural odds) foals in the smaller crop having two of the eight critical genes is 1,680/64, or 26 rounded to the nearest integer. The next higher probability based on multiples of 1/8, 1/512, probably represents an upper limit for Man o' War's crop. It implies that there may have been 1,680/512, or 3 foals in the crop having three critical genes.

Thus, that 26 foals within his crop possessed at least two of the eight critical genes is likely, but it is much less so that any had three such genes, other than Man o' War himself.

For Secretariat, it is likely that 24,361/64, or 381, foals possessed two critical genes that 24,361/512, or 47, had three critical genes and that 24,361/4096, or 6, foals carried 4 such genes.

It is likely that Secretariat carried five or more critical genes. His Triple Crown achievements alone, nearly suggest that the latter possibility was realized.

The presumptions being clearly stated, it still does not mean that a foal having four critical genes will necessarily be a great runner. The genes may not be in the proper combination to manifest as what humans call greatness.

Recall that, of eight genes, the number of possible random combinations having four each is found by the formula n! ÷ [r!·(n-r)!]. In this formula n stands for the total number of items eligible for combination, and r is the number of items in the subgroup to be formed from them.

In this example, n = 8 and r = 4. Expressions like n! and r! are read: 'n-factorial' and 'r-factorial.' The factorial sign means that, to calculate the equivalent value, begin with the number representing the total items being considered, 8 or 4 in this case, then multiply it by successively smaller integers until 1 is reached.

Thus, when n equals 8, n! = 8 x 7 x 6 x . . . x 1 = 40,320; similarly, when r equals 4, r! = 4 x 3 x 2 x 1 = 24. The same procedure is used for the parenthetical expression. Thus, (n - r)! = (8 - 4)! = 4! = 24. The final answer is: 70. Therefore, there are 70 possible combinations of eight genes taken four at a time. Many of these combinations may not produce great or even good runners.

Using the Genetic Arguments for Practical Data Comparisons

When Man o' War's and Secretariat's data are compared in this book, the above genetic discussion is always implicitly invoked. That is, all data adjustments are based on the consequences of arguing from a presumed *constant genetic basis* of Thoroughbred ability as postulated above.

For an example in the time domain, when Man o' War's juvenile times for six furlongs are being adjusted for possible track or impost effects, the suggested adjustment is always examined for consistency with the *assumed constancy of Thoroughbred running ability* across many generations.

Thus, for his six six-furlong juvenile races it is found that Man o' War averaged, without adjustments, 72.21 s with standard deviation 0.72 s. Thus, his minus-three-sigma level, the fastest he would likely have run six furlongs, equals 70.05 s.

However, when this number is compared with the 2011 world record holder for six furlongs, Twin Sparks, his time of 66.49 s (ARM 911) represents a rare -3.43 standard deviations below the *average* for the world's 48 fastest six-furlong times.

Twin Sparks represents a good, but not great, Thoroughbred. His career record included 29 starts with 9 wins, 3 places and 4 shows. He was thus a multiple stakes winner and definitely within that magical 2.5 or 5 percent segment of his foal crop.

He has not, as of this writing at least, been selected as a top champion for inclusion within either edition of the DRF Champions books. He possibly ran the race of his life in setting the six-furlong world dirt record, and it is arguable that Man o' War could also have run at least an equivalent time had track and other conditions permitted. The same, naturally, holds for Secretariat.

One simply cannot believe, unscientific as that statement is, that Man o' War and Secretariat were not at least equivalent runners to Twin Sparks.

Since Man o' War's standard deviation for six furlongs was 0.72, then multiplying that by 3.43 and subtracting from his average time gives 69.74 s. This time, from merely a hasty and superficial adjustment, is assumed to be the *minimal* attainable level for Man o' War running on a track that was undoubtedly slower than that on which Twin Sparks ran in 2009. It adheres to the boundaries suggested by Figure 7 and discussed later.

Therefore, the minimum reasonable time that Man o' War should be considered capable of attaining for six furlongs is *at least* 69.74 s and possibly less. This time thus passes the "Constancy Test" for Thoroughbred running ability across generations and would be a reasonable initial adjustment, possibly amenable to further lowering depending on other evidence.

An analogous calculation for Secretariat, having an average of 70.47 s and standard deviation of 0.42 s for six furlongs gives the minimally reasonable time of 69.03 s.

The above considerations also directly apply to the abilities of the current nine-furlong dirt record holder, Simply Majestic. His career record is a respectable 44 starts with 18 wins, four places and seven shows. Like Twin Sparks, he is not included in the DRF Champions publications, but he set a blistering nine-furlong record time of 1:45.00 at Golden Gate Fields in 1988. (ARM 911) That is the way Thoroughbred ability often works. One superlative effort does not generally bequeath immortality. Ainslie discusses related issues, especially concerning how the herd instinct of Thoroughbreds can affect running. (213)

The ultimate goal of this book is to suggest *tentative* data adjustments, primarily for Man o' War but also occasionally for

Secretariat, which are both practical and realistic. In all cases, a brief discussion of possible data distortion effects will always accompany the results of any given adjustment.

A systematic approach to these adjustments is always preferred. That is, adjustments proceed from what I consider the simplest and most obvious to the more detailed and less obvious.

In this regard, I strive to remember a primary dictum within the field of physics—that the best possible, and most probably correct, explanation for some phenomenon is the most self-evident of any posited explanations. This rule, with several variations, is also called "Ockham's Razor" after the fourteenth-century philosopher and theologian, William of Ockham. (Ziccardi 6)

There undoubtedly exists a *single, true value* of adjustment that would render Man o' War's record *exactly comparable* to Secretariat's. It is a value that would finally close the books of debate regarding which horse was *really* faster.

Whether or not that value is discoverable by standard statistical methods depends greatly upon the imagination and perseverance of the researcher—and probably on a strange factor called serendipity. It does not require invoking mythical effects, such as an era influence, that are never defined.

The possible reward is worth the effort, even though it attracts the ire and contumely of those who never quit their noisome chant: "You can't compare horses from different eras!"

The reader is hereby invited to sit back and watch.

Chance is perhaps the pseudonym of God when He does not wish to sign His work.
~ Anatole France

Chapter One

Introduction - Establishing the Basics

When *Thoroughbred Champions: Top 100 Racehorses of the 20th Century* was published in 1999 by The Blood-Horse, Inc., even one of its voting panel members expressed doubt concerning its meaning (Nack 8).

Following that publication, a certain animosity was activated, or exacerbated as you wish, between advocates of Man o' War and partisans of Secretariat. Internet cites proclaiming it are easily found without specific directions.

These two colts, arguably the greatest North American runners of the past century, and possibly the greatest runners of any century from any country, completed feats and established records of near mythical proportions.

Man o' War was foaled on March 29, 1917 just minutes before midnight. Secretariat's foaling occurred at 12:10 AM on March 30, 1970 (Wolfe 25). Thus, they were historically separated by essentially 53 years to the minute.

This time separation is generally referred to as an 'era difference' by their respective devotees whenever manifest running ability is discussed. More exactly, it is prevalent in such statements as: "You can't compare horses from different eras!" This mantra-like utterance is also what I think of as the 'forbidden comparison.'

Obviously, I do not subscribe to this dogmatic viewpoint. My purpose in writing this book is to *compare* the past performance records of these horses (technically, colts) using unbiased statistical methods. These analyses will be applied systematically using what I believe are the minimum techniques necessary for a thorough assessment.

No statements issue from these analyses regarding the relative quality of either colt. This fact will be stressed repeatedly throughout the book. The idea behind this caveat is that past performance records indicate *what* the particular horse *did* at a given time and place and under specific circumstances.

Past performance records do not indicate what a horse *was capable of doing* given ideal circumstances. Such circumstances seldom hold, even at the most restricted level.

One can, for instance, nearly always find an element of 'bad luck' which diminished a given performance and prevented it from showing the horse's full capability.

Such is the situation in nearly every sporting event for both humans and equines. The world simply is not an ideal place. Vagaries of chance always touch events on some level.

A quotation from Kent Hollingsworth on this very question is appropriate. He was speaking of Colin, who remains one of the few unbeaten Thoroughbred runners of the twentieth century: "Great horses have been beaten by mischance, racing luck, injury and lesser horses running the race of their lives. None of these, however, took Colin. He was unbeatable." (Colin's Ghost 2)

Multiple factors constantly operate within each horse race, and these factors ultimately determine which horse wins. This concept is presented graphically in GHA. (Justice 67)

Chapter One

These factors can be readily examined and compared using basic statistics concepts. Six primary statistical concepts are introduced and fully explained in this book. Readers comfortable with eighth-grade level mathematics will have little or no difficulty following the explanations and examples given herein. Ample appendices support the textual discussions.

It is absolutely necessary to use at least this minimum set of statistical concepts. Objective, scientific comparison of thoroughbred racing data is impossible otherwise. Without using elementary statistics to compare racing data, only subjective impressions, anecdotal comments and other such imprecise evaluations are possible.

Current internet diatribes about whether Man o' War or Secretariat was the greatest horse of all time invariably exclude statistical comparison. Therefore, they are based solely on opinion. I often judge such opinions as prolonged rooftop shouting. Many people apparently think that if you keep shouting something loudly enough and long enough it becomes truth.

Since opinion is generally unreliable, let us proceed to a systematic development of the sole methods which are reliable.

The World of Thoroughbred Racing

First comes the race, then comes the recording of facts about the race, and last comes the analysis and comparison of those recorded facts. The basics of the race will be discussed first, in keeping with the idea of a systematic approach. No pre-existing knowledge is assumed on the reader's part regarding the elementary foundation of either horse racing itself or of statistical analysis applied to racing data.

By following this approach, I hope to make explicit and clearly understood what can otherwise become thorny questions and muddled conclusions.

Background Basics
Distance

Three elements comprise the crucial data sources for a horse race. They are the same as the primary units of measure used in elementary physical science. These units are *distance*, *time* and *mass*. Two secondary units, speed (velocity) and acceleration are also used as derivatives of the primary units.

The furlong, abbreviation f, is the elementary unit of distance in horse racing. Its name is said to derive from the middle-English word for 'plowed furrow.' The furlong is defined and set at exactly one-eighth of a statute mile, or 660.00 feet. The statute mile is 5,280.00 feet. Hereafter, for brevity, this distance and its unit are expressed as: 5,280 ft.

One can get a visual idea of a furlong by considering that it is two and one-fifth times the distance from one goal line to the other on a standard American football field.

In this book, furlongs and miles, plus fractional miles, are both used as general distance descriptors. For calculations, however, race distances are converted to feet because this makes more sense, both computationally and cognitively.

For example, a one-mile race is identical to eight furlongs or 8 f. This distance is 5,280 ft – the formal, two-decimal-place zeros generally being dropped for such generalized discussion.

Note that 5,280 ft is the *nominal* distance—the stated distance—of the race being considered. In most cases, horses in a one-mile nominal race actually travel further than one mile since

Chapter One

they must often be "wide of" the rail to negotiate around other horses.

The particular path a given horse takes in running and completing a given race is called its 'trip.' And that is, perhaps, the extent of the humor used among racing aficionados!

The nominal distance is always as measured around the track from start to finish and *immediately adjacent to* the rail. This fact will especially be noted later when several of Secretariat's major races are discussed.

Time

Entire volumes have been written about time. Most people probably consider time, if they even consider such concepts, as an actual entity of some kind that exists "outside" the human mind. Time is then what is "measured" by watches or clocks or, more academically, by chronometers – just as lengths are measured by rulers.

The great scientist, Sir Isaac Newton, thought that time existed in this universal and absolute sense, outside the mind. Many current scientists do not agree. In fact, Albert Einstein went so far as to say that space and time, formerly considered as separate entities, are actually combined in a space-time continuum.

This is obviously not the place for further elaboration, but it may be a point of interest for some readers.

Whatever time actually is, it is generally said to be measured, or indicated at least, by watches in elementary units called 'seconds.' There are 86,400 of these seconds in one 24-hour period.

This period is generally called one day, although the earth does not actually rotate in exactly this time. The actual rotation value of earth relative to the stars is: 23 hours, 56 minutes and 4.091 seconds. (Time Almanac 399)

However, using the accepted 86,400 value, then a horse race taking exactly 2 minutes or 120 seconds consumes 120/86,400 or 0.00139 of a day – to five decimal places.

That time happens to be an extremely fast one for a race like the Kentucky Derby which is 10 f or 1.25 miles. In fact, the Derby record, set by Secretariat in 1973 and still holding, is 119.40 seconds.

Mass versus Weight: Other Names for Impost

The topic of mass is important in Thoroughbred racing because we will later calculate the kinetic energies of horses, and mass is used in that context. These energies are directly relatable, within limits, to the weights or impost carried by a horse in a given race. More important, such energies are indicators of what is called the 'speed' of a given racetrack.

Until now the concept of track speed has generally remained ambiguous. It has not been placed on an objective footing. Using kinetic energy allows this to be approximated.

There is nothing mysterious about the mass of an object. The objects whose mass are of interest in a horse race are the jockey and the tack (saddle, blanket, etc.) that the horse must carry around the track. The weight or mass of a horse's racing plates (shoes) is also a factor, and it will be discussed later when such refinements are necessary to make a point.

To calculate the mass that the horse carries divide the total weight or impost, expressed in pounds, by 32.2 feet-per-second squared (abbreviated ft/s^2 or $ft\text{-}s^{-2}$), the acceleration of gravity near the earth's surface. This notation is explained below after average speed is discussed.

Chapter One

Average Speed

If we are given that a horse has run one mile, 5,280 ft, in 95.00 seconds, then his *average speed* is found from the formula $d = \bar{s} \cdot t$, where the dot, '•', means to multiply the values for '$\bar{s}$' and 't' together. Using the dot allows clearer reading of text than using 'x,' which is often used as the universal sign for multiplication.

In the formula 'd' represents the race distance and 't' represents the time for the race.

Since we are given the distance of the race and the time in which it was run, dividing distance by time will give the average speed, $\bar{s}$, for the race. The lower-case 's' with a macron above it is the generally accepted symbol for average speed.

In many cases, the symbol $\bar{v}$ will be used, especially when kinetic energy is discussed later. This symbol means average velocity. Speed and velocity, and their average values, are directly related. They are actually interchangeable in terms of horse racing data.

If you are told that someone is traveling 60 miles per hour, then his *speed* is being given. If you are told that someone is traveling 60 miles per hour *in the direction* north-northwest, then his *velocity* is being given.

This may sound like a rather minor distinction, but it is important in many science applications. For horse racing, speed provides complete information about the horse's ability to cover ground in a certain time, and velocities are not needed.

Speed is called a *scalar* quantity. Velocity is called a *vector* quantity. A scalar quantity has only *magnitude* associated with it, whereas a vector quantity has both *magnitude* and *direction*.

Acceleration

Acceleration is a vector quantity just as is velocity. If an object is being accelerated, either its speed or its direction, or both, are changing. If the change in speed remains constant for a given time interval, then we say that the body has undergone constant acceleration.

The acceleration due to gravity, discussed previously, is an example of a constant acceleration, provided the object to which it is applied does not stray far from the earth's surface. The magnitude of the acceleration due to gravity, g, is generally stated as 32.2 ft/s^2 (read: feet-per-second squared) at or near sea level over the entire surface of earth. The *direction* of earth's gravity is always taken as directed *downward* precisely toward the earth's center.

The notation, ft/s^2 can be baffling. It may be read as either 'feet-per-second per second, or 'feet-per-second squared.' The expressions sound awkward. They just mean that an object's speed is increasing at a rate of so many feet per second *for each second* it moves – hence, the ingenious term "units-per-second squared."

An equivalent notation for the same statement is ft-s^{-2}. This notation is read exactly as the former and means the same thing. It simply allows placing the 's' on the same line with 'ft,' because the '-2' exponent actually means to divide by s-squared.

An example will clarify how acceleration is interpreted generally.

Suppose we are given that a certain horse left the starting gate and reached the first half-furlong point of a race in 6 seconds. What was his acceleration?

Chapter One

Recall, that one furlong represents 660 ft. Therefore, a half furlong is 330 ft. His average acceleration, $\bar{a}$, is found by dividing the change in his average velocity *by the time it took for that change to occur*. This is symbolized by $\Delta \bar{v}/\Delta t$ and means 'change in average velocity (or equivalent speed) *divided by* the change in the corresponding time interval. The Greek capital letter delta, Δ, is generally used in mathematics to denote change.

The speed, or velocity, at the instant the horse left the starting gate was zero. His *average velocity* at one-half furlong equals the magnitude of his average speed for that distance. The speed is found from the formula $d = \bar{s} \cdot t$. Rearranging this equation to solve for average speed gives $\bar{s} = d \div t = 330$ ft $\div 6$ s $= 55$ ft/s.

The average acceleration is, therefore, the final average speed minus the initial speed (55 ft/s – 0 ft/s) divided by the corresponding time interval: 55 ft/s $\div$ 6 s, or $\bar{a} = 9.17$ ft/s^2. The result is read: "9-point-17-feet-per-second squared." It means, as previously stated, that the horse accelerated or changed speed by an average of 9.17 feet per second *for each second* he ran.

Thus, during the first second, his average speed changed from zero at the starting gate to 9.17 ft/s. After 2 seconds his average speed was 9.17 ft/s + 9.17 ft/s or 18.34 ft/s, since by definition he is accelerating 9.17 ft/s for each second he runs.

After 3, 4, 5 and 6 seconds, his average speeds were, respectively, 27.51 ft/s, 36.68 ft/s, 45.85 ft/s and 55.02 ft/s. The '.02' in the final number should be exactly '.00,' but rounding error makes this negligible difference.

The important point to take from this example is that you understand what acceleration indicates – namely, how fast the speed of an object is changing per unit time.

Momentum and Kinetic Energy

Momentum and Kinetic Energy are the only remaining parameters to consider regarding the movements of racehorses. The remaining items are statistical in nature and are described in the next chapter.

Momentum is defined by the equation $p = m \cdot v$. An object's momentum, p, is just the magnitude of its mass multiplied by the magnitude of its speed or velocity.

An object's kinetic energy is defined by the equation $KE = \frac{1}{2} \cdot m \cdot v^2$. This equation means to multiply the fraction ½ by the object's mass and then by the square of its velocity or speed to obtain the magnitude of its kinetic energy.

Both parameters are discussed later when the effects on the horse of impost and track speed are considered. Since you are now familiar with the concepts of mass and velocity, or speed, there should be little confusion about what the two preceding equations mean.

Having established the basic vocabulary and parameters directly related to the race track and the race proper, we turn to the basic statistics which are applied to compare past performance records and make judgments concerning them.

If you had a complex task to do and needed help, whom would you rather have for assistance—two people with IQ 70 each or one person with IQ 140?
~ Walter Sullins

Chapter Two

The Statistician's World
Populations and Samples

Numbers form the heart of statistics. Statisticians generally recognize four classes of numbers. This book uses only those numbers that can be placed within the scale class called a "ratio scale." Such numbers and their scale allow direct comparison between any two numbers on the scale. They are considered the numbers and scale for all scientific measurements. And, by definition, they have a true zero point.

For example, contrary to the numbers alluded to in this chapter's leading quotation by my late, esteemed statistics professor at Indiana State University, lengths are measured in numbers which logically fall along a ratio scale, as does time.

If we say that one table is 48 inches long and that another is 24 inches long, then it is perfectly logical to say that the former table is exactly twice as long as the latter, or that the latter is one-half as long as the former.

If we say it takes six hours to make one drive and only two hours to make another, then nobody doubts that the former drive is three times longer, in time at least, than the second.

A ratio scale has a *true zero point*, thus allowing meaningful ratios to be declared between all other points along it. Scarcely anyone should be surprised by that fact. Not all numbers, those representing IQ included, are locatable along such a scale – hence, the quotation's humor.

Fortunately, distance and time, the two major parameters of horse racing compared herein, fall naturally on a ratio scale. Thus, they evade Dr. Sullins' pointed query.

Statistics is the set of mathematical techniques one uses to find trends, patterns or relationships within groups of numbers which, preferably, fall on ratio scales. These numbers come from two groups, *populations* and *samples*.

Populations are usually larger groups of items (events, people, inanimate objects, etc.) from which much smaller groups, called samples, are randomly selected.

The samples are then used in an attempt to accurately describe *parameters* of the population. These parameters correspond to related sample values called *statistics*.

Strictly speaking, the term 'statistics' is ascribed both to the general mathematical study of patterns within data and to specific calculations made on sample items.

Samples are used because they are generally smaller than and thus faster, cheaper and much less time consuming to study, than are entire populations.

For example, one may wish to determine the overall IQ of the senior class at a certain high school. Assume the class has 200 members. Rather than attempt to schedule and conduct IQ tests on all 200 members, a random sample of, say, 30 seniors

Chapter Two

may be drawn and the same test administered to each member of this much smaller group.

Sampling theory says that a group of 30 or more is generally sufficient from which to draw valid conclusions about the corresponding population parameter (in this case average IQ) using the smaller group, or sample, value.

Many statistics texts cite the number '30' as being the size needed for a statistically significant (large) sample, but they offer no real explanation. At least one clear and succinct explanation is available on the Internet. (Campbell 1)

The above example of population and sample was straightforward. However, defining populations can be ambiguous depending on the intentions of the investigator.

For instance, one text (Downing 19) gives an example of traveling across the country and crossing seven named suspension bridges. It also lists the lengths of the seven bridges and then shows how to calculate the mean, variance and standard deviation on those lengths.

The confusing part of the example is the implication that the means and variances described were calculated on a population, whereas a sample mean or variance is calculated slightly differently.

If the authors truly interpreted the set of seven suspension bridges as a population, then I do not understand their reasoning. It can be easily verified, via the internet for example, that 79 active suspension bridges currently exist within the United States. The dates for those bridges show that they existed prior to the latest publication date, 2009, of the book. Therefore, the authors should have been aware that the particular set of seven bridges used for illustration was not the natural population of such bridges.

Clearly, the set of seven bridges used for demonstration constitute a sample of those 79 bridges rather than being the entire plausible population.

The fact that population definitions can sometimes present interpretation difficulties will be addressed again when linear regression is later discussed.

Descriptive and Inferential Statistics

Two general types of statistics exist: *descriptive* and *inferential*. Descriptive statistics include the *average* or *mean* of a *set* of data, the *standard deviation* and *variance* of that set and also its *range*. There are other parameters involved in descriptive statistics, but these four are sufficient for comparing horse racing data.

Inferential statistics is concerned with stating hypotheses, and then rejecting or not rejecting them based on certain calculated values, about the populations from which the samples came.

Note that *samples should always be drawn randomly* from a given population. Only in this way can one be reasonably certain that bias does not affect the data. Randomness assures that *each item* in the population *has an equal chance* of being selected for the sample.

Various ways of random selection are available, from tables of random numbers all the way down to flipping a coin and making a sample selection based on the outcome. This book uses Excel's Random Number Generator (RNG) function exclusively.

We now discuss the four principal statistics used to describe sample data specifically for analyzing racehorse performance: the mean, the range, the standard deviation and the variance.

Chapter Two

Mean or Average of a Sample

A *set* or *group* of sample data can generally be denoted by a sequence of numbers within braces. An example is {8, 10, 9, 9, 7, 9, 11, 12, 11, 10}. The numbers represent individual measurements made on some characteristic of objects or events we wish to study.

The average for the above set is found by simply adding the ten items of the set and then dividing that total by the number of items added, 10.

The Greek capital 'S,' or 'Σ' (sigma), is generally used in statistics to denote a sum of several items. Therefore, the average of the above set is found by calculating:

$$\bar{x} = \Sigma\,(8+10+9+9+7+9+11+12+11+10) \div 10 = 96 \div 10 = 9.6$$

The symbol $\bar{x}$ (called 'x-bar') is nearly a universally accepted symbol for the average or mean of a set of data.

Range

The range of the above data set is simply found by subtracting its smallest data value from its largest data value. In this case the range is: $R = 12 - 7 = 5$.

Standard Deviation and Variance

The standard deviation (SD) is a kind of data average. It describes how far the data spread out on both sides of the mean or average. Rather than clutter the text at this point, the method for calculating the standard deviation of any set of data is completely shown in Appendix B.

Appendix B also shows how the variance is calculated once the standard deviation is known. Calculating variance is very easy because variance is simply the square of the standard deviation. The variance becomes important when the t-test is described later in this chapter.

Microsoft Excel© is a very good statistical applications package and easily calculates all the statistics described and used herein.

The Normal Distribution

The normal distribution is also variously called the normal curve, the bell curve, and the Gaussian distribution. This curve is undoubtedly the most important single distribution within statistics, although it is not the only distribution used.

The normal distribution has an exact mathematical equation which will not be described here because knowledge of it is not needed for applying the curve. The equation was first derived by Abraham DeMoivre in 1733. (Walpole 129) Because the German mathematician, Carl Friedrich Gauss, was more famous and developed applications for it, the normal curve became known as the Gaussian distribution.

What is a distribution? It is basically a curved line on a graph showing how the data in a sample are spread out—or *distributed*. The curve's height shows the relative number of sample values within a given interval of the total range of values.

For the normal distribution, the larger portion of sample data is located near the distribution's average or mean, directly beneath the curve's peak, and the number of data items becomes gradually smaller the farther from the mean one goes—outward on either side of the mean toward the distribution's 'tails,' as they are called.

Figure 1 is of a generic normal distribution generated in Excel by using Schmuller's method (121).

Figure 1
Normal Distribution with Mean 10
and Standard Deviation 2

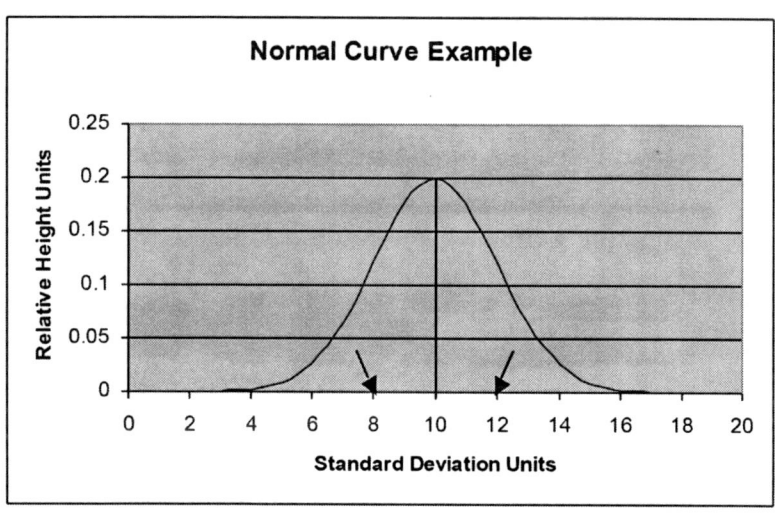

The *standard normal distribution* has, by definition, a mean of zero and a standard deviation of 1. The curve in Figure 1 shows its standard deviation, 2 units, going outward in both directions from the mean. The arrows indicate − 1 SD (abscissa value 8) left of the mean and + 1 SD (abscissa value 12) to the right of the mean.

The figure might suggest the conclusion that the curve ends at about the third standard deviation (abscissa values 4 and 16) on either side of the mean, since the normal distribution is symmetrical about its mean. However, the curve theoretically goes to infinity in both directions without ever touching the horizontal axis. The curve is, therefore, said to be *asymptotic* to the horizontal axis in both directions.

For practical purposes, one need not consider the areas beyond ± (plus or minus) 3 standard deviations (3σ) on either side of the mean. These limits are usually called the 'three-sigma' limits. The total area under the distribution's two tails and beyond ± 3σ is 0.27 percent of the entire area under the curve which, by mathematical definition, is 1.00.

Important, specific areas exist under the normal curve between *each paired set* of standard deviations on *both sides* of the mean. These areas are extremely useful and important when comparing simulated race results. They are given here for reference. All ± SD points are symmetrically placed on both sides of the mean in Figure 1

Between the mean and ± 1 SD on *either* side, the fractional area is 0.3413 of the total area under the curve (34.13%). *Between* ± 1 and ± 2 SD on *either side* of the mean, the area is 0.1359 (13.59%) of the total. Last, for useful purposes, between ± 2 and ± 3 SD on either side of the mean, the area is 0.0215 (2.15%) of the total.

Adding these three areas and then subtracting from 0.5000 (50%), the total area in both the *left and right halves* under the curve, gives approximately 0.0013 as the area remaining under *each tail* of the normal distribution.

This small area is insignificant when considering the results of simulating races between Thoroughbreds. More will be discussed later on this topic when simulation is considered.

Chapter Two

The normal curve has been long known to describe many distributions that occur naturally. (Denny 22) This is why it has come to assume probably the greatest significance of any distribution within the study of statistics. Most of the Thoroughbred data discussed herein were tested using the Shapiro-Wilk criterion for normalcy and found to be, within acceptable error limits, normally distributed.

Correlation

In statistics, correlation places a numerical value on the *strength of the relationship* between two or more data sets. The values assigned are always between -1.00 and + 1.00.

Correlation, by itself, *does not imply* that two or more things, items or events, whose measured values comprise the sets, are causally related—that is, that one *causes* another. The items may or may not be thus related, but additional facts beyond standard statistics are generally needed to show that relationship.

This is a crucial point that many users of statistics are either unaware of or ignore.

The following three sets of paired numbers provide examples of correlations.

Correlation = + 1.00		Correlation = - 1.00		Correlation = -0.76	
1	2	1	16	3	4
3	4	3	14	5	3
5	6	5	12	1	5
7	8	7	10	1	4
9	10	9	8	5	3
11	12	11	6	3	3
13	14	13	4	4	3
15	16	15	2	4	2

As the above data show, a correlation of + 1.00 implies that two sets of numbers remain in 'perfect step' with each other. They need not be in exactly the same proportions as used here for an example, but as one set increases, the other set always increases.

For a correlation of – 1.00, as one set of numbers changes the other changes in the opposite direction. That is, as the numbers in one set become larger, the numbers in the second set become progressively smaller, or vice versa.

The third pair of example numbers was generated by rolling two dice. The numbers of the leftmost set represent the values indicated on the first die; the numbers of the rightmost set show the corresponding values obtained on the second die.

Figure 2 is a graph of this relationship. It serves to provide visual reinforcement of what correlation indicates.

Note that the numbers corresponding to the leftmost set of die throws are plotted along the horizontal axis. This set is expressed as: {1, 3, 4, 5}. The numbers corresponding to the rightmost set of die throws are plotted along the vertical axis. This set is expressed as: {2, 3, 4, 5}.

The single data point, showing 5 on the horizontal axis and 3 on the vertical axis is actually a double point. Both times a five was rolled on the first die a three happened to show on the second die. Since the values are superimposed on the graph, it makes it appear that only one point rather than two is located in this position.

Chapter Two

Figure 2

Plot of paired numbers obtained by dice rolls

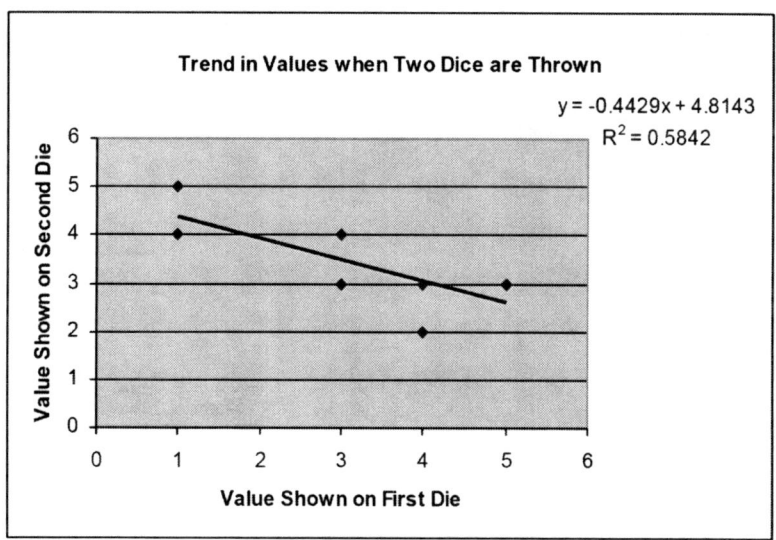

The graph axes, the first going horizontally across the page and the second going vertically up the page, are traditionally called the *abscissa* and the *ordinate*.

The variable called the 'independent' variable is always plotted along the abscissa, and the variable called the 'dependent' variable is always plotted along the ordinate.

For this example, as opposed to the data plotted for horse racing, only *integer values* can occur on either axis because the results of any die throw are "quantized," to borrow an expression from physics. That is, the values can only be integer numbers because those are the only results obtainable on the six faces of a die.

Notice that the value of R^2, the *coefficient of determination or COD,* associated with the graph and shown near the upper right corner, indicates that about 58% of the change in the second die's value can be explained by changes in the first die's value.

This presents an anomaly only explainable by the small sample size. In this case, the sample size is 8, the total number of dice throws.

Many basic statistics texts use throwing dice as an example of *independent events*. That is, the outcome of one event, throwing the first die, presumably has no influence whatsoever on the outcome of the second event, the value that comes up on the second die. In that case, however, the correlation between the numbers representing the results of each throw should be zero or reasonably close to zero.

Again, we can assume that the vagaries of chance acting on this small sample size are the reason for this "false" and inflated correlation.

Note also that the line on the graph which approximates the trend of the points slants downward from the upper left toward the lower right of the graph.

Whenever two variables are correlated negatively, a line representing that correlation will always slope in this direction. For positive correlations, such as those you will always have when a horse's times are plotted (vertically)versus the distances (horizontally) he ran, the resulting line will *always* slant upward from lower left toward the upper right-hand corner of the graph.

An actual "perfect" random correlation would present a horizontal line and a correlation of exactly 0.00.

The amount of the *slope*, either positive or negative, is shown by the number that multiplies 'x' in the graph's equation. This number, -0.4429 in Figure 2, is called the *coefficient* of 'x.' It tells how rapidly the trend line falls or rises for each unit increase along the horizontal axis. Since the coefficient is negative in this case, the line slopes downward.

The last number of interest in the graph, 4.8143, is the *intercept*. It indicates the point on the vertical axis where the trend line would intersect it, were it extended backward far enough.

By substituting the numeral zero in the equation for the value of 'x,' the result $\hat{y} = 0 + 4.8143$ is obtained. This indicates that when x = 0, $\hat{y}$ = 4.8143, which is the y-intercept of the line. The accepted symbol for a predicted value is '$\hat{y}$.'

Without actually projecting the line backwards, it is visually reasonable from the graph that so doing would result in its crossing the y-axis near the value 4.8.

While the y-intercept is meaningful in some cases, it is not in this case. Always recall what the numbers on the two axes represent when a question of meaning arises.

In this example the axis numbers represent possible values that show on a pair of thrown dice. Since neither zero nor 4.8 can occur on a dice throw, then the intercept in this case is just an artifact of the way the computer fits the line to the data.

There is a mathematical equation the computer follows to calculate the correlation coefficient, r, between two sets of data. That equation is given in Appendix C. It would be beneficial if you worked through that example to convince yourself that you understand the basic process of the calculation. It is not difficult.

As a final comment before proceeding to linear regression, a second set of throws for a pair of dice was done just to see what the new correlation would be. The pairs of values obtained on eight throws are given by the set: {3/2, 3/3, 5/4, 1/4, 4/6, 6/2, 6/6, 6/5}.

Using braces, { }, is the general mathematical way of showing a set of values. In this case all entries are given in the form a/b, where 'a' is the value on the first die and 'b' is the value on the second die.

Use these eight pairs of values and work the correlation formula to convince yourself that you can get the proper result. Check your answer by entering the pairs of numbers into an Excel spreadsheet so that the first value of each pair is in one column and the second value is in a corresponding adjacent column to its right.

After the pairs of values are entered, first select a nearby cell and then left-click the cursor on the Insert Function button 'f_x' near the top of Excel's screen. Select 'CORREL' from the 'Insert Function' menu that opens. A Function Arguments window (second window) then opens. In the 'Array 1' field of that window enter the first set of values by *first clicking once in the Array 1 field window* and *then* clicking and dragging the cursor over the *leftmost column of values* entered. Then click in the 'Array 2' field of the window and then click and drag the cursor over the *second set of values in the adjacent, rightmost column.*

End by clicking 'OK' and the correlation between the two sets of values will appear in the initial cell selected. This should match the value calculated by hand using the formula from Appendix C.

It will also be helpful to work through the example in Appendix C regarding the manual calculation of correlation coefficients.

The formal name for this method of calculating correlation coefficients is the Pearson Product-Moment Correlation

Coefficient, after its developer, psychologist Karl Pearson. See, for example, texts such as Ferguson (104).

You now have enough information to understand a key concept used in statistics—simple linear regression. This involves how a so-called trend line, as in Figure 2, is actually adjusted for the best "fit" to a set of data points.

Simple Linear Regression

Simple linear regression (SLR) involves two variables, just as were used in the dice throwing example. For horse racing data, the *distance* of each race a horse runs is designated the *independent variable*, and the *time* for running each race is the *dependent variable*.

Distance is an independent variable because it can be controlled or set by the people establishing the race requirements. The running time, however, cannot be controlled. Only the quality of the winning horse, the jockey's skill and the particular race conditions under which the horse runs determine the winning time.

Therefore, distance is plotted along the horizontal axis, the abscissa, and time is plotted along the vertical axis, the ordinate.

Refer now to Figure 3.

Figure 3
Example of Simple Linear Regression

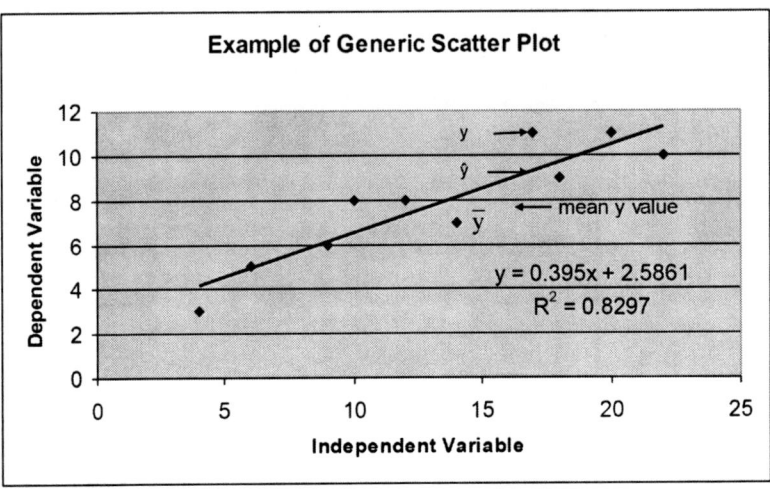

Ten data points are plotted on the graph, and the linear regression line is shown running diagonally upward from left to right among those points.

Each point is associated with a pair of numbers (coordinates) that uniquely determines its placement on the graph. These coordinate numbers are always given in the order (x, y) when locating a point. The 'x' refers to the point's distance along the x axis from the graph's origin, 0, and the 'y' refers to the point's distance, above or below the x-axis, depending on the ordinate's sign.

All points *right* of the origin on the x axis are considered *positive*, as are points *upward* along the y axis. Points in the opposite directions along either axis are considered *negative*.

The leftmost point in Figure 3, just beneath the left end of the regression line, for example, has coordinates x = 4, y = 3. The rightmost point, near the upper right-hand corner of the regression line, has coordinates x = 22, y = 10.

The regression line itself is "fit" by the computer's Excel application so that the sum of the *squared vertical distances* of *each* of the ten data points *from* the point on the regression line either directly above or below it is a minimum.

Ten such distances are calculated for this particular graph since there are ten data points. Each individual distance is represented by $y - \hat{y}$. Statisticians nearly always use the symbol $\hat{y}$ to represent a *predicted value* obtained from a regression equation. Excel does not print $\hat{y}$ on charts.

The *difference* between any observed data point and the corresponding predicted point of the regression line, $y - \hat{y}$, is termed a *residual*. It can be interpreted as the amount left over or unexplained by the regression line. If the regression line perfectly predicted the location of the given data point, the residual would be zero. This is an important concept. The mean of all ten y values, 7.8, is indicated by $\bar{y}$, but the numeral is intentionally omitted on the graph for simplicity.

Refer to any basic statistics text, such as Schmuller (223) or Ferguson (114) for a discussion of linear regression and the *method of least squares*, as the fitting technique is called.

When Excel's LINEST application is run on a paired set of data points such as those in Figure 3, it produces a small output table (an array) having two columns and five rows of cells. Appendix D shows the LINEST output table or matrix with explanations of the ten numbers it contains.

The four numbers of immediate interest for interpreting race results are highlighted in the table. Three of them appear in the equation Figure 3 displays in its lower right-hand corner. These four numbers include the *slope* of the regression line, 0.395, also the coefficient of the independent variable x, the regression line's *intercept*, 2.5861, where it would cross the y axis if extended backwards, the *coefficient of determination*, 0.8297, or R^2, and the standard *error of estimate*, 1.144978, the *estimated* population standard deviation for each predicted point, ŷ, if multiple, random linear regressions were run.

Walpole and Myers (331) include an extended discussion of the standard error of estimate, as used in prediction, but it is on a moderately difficult level. However, they provide excellent diagrams to support their textual explanation.

The slope indicates that, for *each unit increase* in x along its axis, the regression line rises (in this case) by 0.395 units parallel to the y axis. Slopes may also be negative, in which case the slope tells how far downward the line drops per unit x increase.

The value of R^2 tells what *percent of change* in the dependent variable is actually *predicted* correctly by changes in the independent variable. In this case, when you multiply 0.8297 by 100 to convert it to a percent, you see that 82.97 % change in the y variable is predicted by changes in the x variable. This point is extremely important.

When you begin studying actual past performance charts of Thoroughbreds, you will find it commonplace that typical regression equations accurately predict 99% or more of time changes based on the distances run. It is a remarkable and extremely useful fact.

This is why, when they are used judiciously, Thoroughbred linear trend lines are as reliable a basis as any for making successful wagers. They are less confusing to use than Beyer Speed Figures and are at least as accurate a method for handicapping.

The foregoing discussion of linear regression should be adequate preparation for those readers wanting to pursue linear regression investigations.

The final major topic in this preparatory chapter, the t-test, is now described. Several minor topics are discussed later within the text as appropriate. They need no actual elaboration now.

Do Two Samples Differ Significantly—Student's t-Test

The concepts of *distribution normalcy* and *random sampling* form the crux of accurate, unbiased Thoroughbred data comparisons.

The reasons are that normal distributions have precise, expected proportions of their area within specified boundaries under their curves and that random sampling helps assure each data point has an equal chance of being selected for a given sample.

Figure 4 displays a typical normal distribution. The box within the distribution represents an area from which a typical random sample might be taken.

Note that the box's height is unimportant. The height was merely selected to emphasize that a given area under the curve contained all the points of some random sample.

It is the box's base *width* that is important because it shows the actual *range of the data within the random sample*. The height only indicates the relative values of *how many points* beneath the curve have a certain location on the horizontal axis.

Figure 4

Normal Distribution with Random Sample Area Indicated

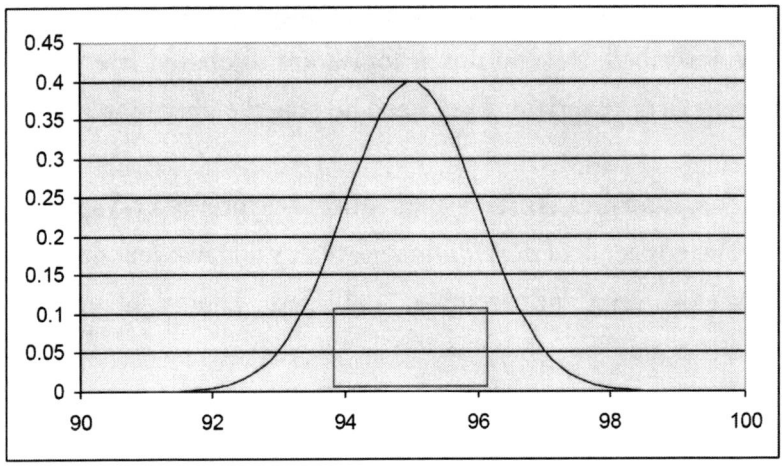

If a *sample* taken *from the population represented by this distribution* were truly random, then one would expect a range of sample values similar to that represented by the box's lower border, extending from about 93.75 to 96.25 along the horizontal axis. This represents the most likely area under the curve where a random sample would originate because the curve is wider in this range than it is between 96 and 98, for example. It is also higher in this region, which means more sample values are likely here.

The t-test is used for small samples—those having less than 30 data values. The reasons for its use are more technical than merit discussion here, but statistics texts such as Ferguson (159), Schmuller (162), or Walpole (259) provide those explanations.

How the t-Test is used

Suppose you have two data samples, one from each of two different horses. Both samples are their times for six-furlong races at the same age. Assume that horse A ran six such races while horse B ran three. It is unnecessary for the number of races in the two samples to be equal to perform the t-test, but a minimum of two data points per sample is needed for running the test.

The t-test is based on a straightforward mathematical formula. The formula represents a fraction having the difference of the sample means and population means in its numerator and the product of the samples' variances in its denominator. Thoroughbred data seldom disclose the population means, but statistical assumptions can still be made about them.

One *always assumes* that samples from *two hypothetical populations* are being compared when the t-test is used. If the test indicates an *insignificant value* for t, this implies that the populations overlapped (their means were fairly close) enough that the random samples could be judged *likely* (having high probability) to have come from either population or from one common population.

If the test indicates it was *unlikely* for random samples to have come from equivalent populations, the test gives an alpha value associated with low probability.

Tables of selected t values are always provided in elementary statistics texts. Appendix E of this book gives an abbreviated t table for reference.

Statisticians generally use a probability level of $\alpha = 0.05$ (five chances in 100) as the typical *lowest probability allowable* for two random samples to have come from populations having equal means. If the *probability* of obtaining a difference between means *as great as, or greater than,* the one obtained by random sampling is *low*, then the difference is termed 'significant,' and it is considered that the samples do not represent equivalent populations. I am using the term 'equivalent' more loosely than most statisticians, but this is not intended as a formal statistics text, even on a beginning level.

This is a somewhat lengthy way of explaining the so-called null hypothesis, which the experimenter must declare or at least decide mentally on, before performing the t-test. The null hypothesis is symbolized by H_0. In words, it says that there is *no difference* (null difference) between the means of the hypothetical populations from which the samples came. That is, $\mu_1 - \mu_2 = 0$, where population means are designated by lower-case Greek letter 'mu.'

The general (unequal variance) formula for calculating t is:

$$t = [(\bar{x}_1 - \bar{x}_2) - (\mu_1 - \mu_2)] \div \sqrt{(s_1^2/n_1) + (s_2^2/n_2)}.$$

In this formula, the entire term within the brackets, [ ], must be divided by the entire expression following the radical sign, $\sqrt{}$. See Schmuller (164-165)

Since the difference between the population means is generally assumed to be zero—that is, the populations are considered identical—then $(\mu_1 - \mu_2)$ is zero, and that part of the formula is basically ignored.

The t-test formula then essentially resolves into comparing the difference between sample means $(\bar{x}_1 - \bar{x}_2)$ to determine whether a significant difference exists between them, and assuming that the corresponding samples are from populations with equal means.

In this case it means we are asking whether or not there is a significant difference between the means of the running times for six furlongs of the two horses, A and B.

The calculated difference in the means is divided by the square root of the entire expression following the division symbol. In that expression, n_1 and n_2 are the sample sizes, 6 and 3 in this example, while s_1^2 and s_2^2 are the variances of the two samples.

The nice thing about the t test is that Excel does this entire calculation. After the data are entered in *two adjacent columns* of the Excel spreadsheet, select a cell for the t-test result and then select f_x from the formula bar; choose 'TTEST' from the 'Insert Function' drop-down menu.

A second drop-down menu labeled 'Function Arguments appears.' Four fields must be completed in that menu. They are named, in top-down order: *Array 1*, *Array 2*, *Tails*, and *Type*.

Click first in the Array 1 field. Then click and drag the mouse cursor over the leftmost column of data. Click the Array 2 field and perform the click-and-drag operation over the cells in the rightmost data column.

Now enter either '1' or '2' in the Tails field, depending, respectively, on whether you judge that one sample contains consistently larger or smaller values than the other or whether you can't tell by direct inspection. Entering '1' sets up a one-tailed t-test; entering '2' sets up the 'two-tailed' test. These are explained in Appendix F.

Last, enter '2' in the 'Type' field if the two sample standard deviations do not test significantly different using the F test explained in Appendix E.

33

If the F test indicates there is no significant difference between the samples' standard deviations, enter '2' in the Type field. Then select the 'OK' option, and the t-test result (α value) appears in the preselected cell. If $\alpha \leq 0.05$, the result is significant.

Note that the result given by TTEST is the *probability value* that the tested samples came from populations having equal means. It is always between 0 and 1. The actual value of t that gives this result is not shown by Excel. The TINV function is used to find the actual value of t that produces the given probability value α. It is unnecessary to find the t value unless you expressly wish to do so.

If two or more samples have approximately equal data spreads—generally resulting in roughly equal standard deviations, this gives rise to what has to be one of the great words within the subject of statistics—*homoscedasticity*! It's quite an academic-sounding mouthful. All it means, however, is that the data spread, or range, within both samples is about equal—that is, they have equal standard deviations.

Hypotheses and the t-Test

The *null hypothesis*, H_0, was already mentioned. That hypothesis is always the first one formally stated before running the t-test. It basically reminds the investigator what he or she is looking for in the data sample comparison—namely, whether or not the difference between the sample means can or cannot be explained just by the nature of random sampling.

The greater the difference between the sample means, the less likely that difference can be explained by random sampling. That is the key point to remember about hypothesis testing. And, the *greater the difference* between the sample means, the *more likely* it is that the t-test will produce an α *below* 0.05. That is the signal that a *significant difference exists* between the sample means.

A second hypothesis is always stated. It complements the null hypothesis. The second hypothesis is called the *alternate hypothesis* and is denoted H_1. The rule for interpreting the t-test is that you have two chances of error when you apply it.

The first error is called *alpha* error. It is denoted by the Greek lower-case letter alpha, 'α.' The second type of possible error is called, perhaps not surprisingly, *beta* error. It is denoted by the Greek lower-case letter beta, 'β.'

The rule for interpreting t-test results is as follows: If you *reject* the null hypothesis when it actually *should be accepted*, this is an example of alpha error. It is also called 'Type I' error. On the other hand, if you *accept* the null hypothesis when it *should have been rejected* in favor of the alternate hypothesis, this is an instance of beta error, also called 'Type II' error.

Figure 5 gives a geometric interpretation of why it is possible for the t-test to lead one into making either of these errors.

The actual difference between the means of populations A and B, which you do not know, is the physical or numerical distance between their peaks, A and B.

If the random samples from populations A and B were drawn from the regions labeled 'α,' *something you also do not know*, the t-test measures this distance as *greater than* the actual distance between the means of A and B since the means of the data points within those regions can be seen to be geometrically within a *narrower range* of the axis.

Since this *measured distance* is larger than the difference between the actual means of samples A and B, this translates into the t-test indicating that the population means are farther apart than they actually are, and so the null hypothesis is rejected. This presents an alpha or Type I error.

Conversely, if the random samples happen to fall within the two regions labeled 'β,' – once again, something you don't know—then the t-test measures that difference in means as *smaller than* the actual difference between the peaks of the population means of A and B.

This, in turn, can cause the t-test to indicate there is no significant difference between the means of the populations when there actually is. Therefore, a Type II or beta error has occurred and the null hypothesis is accepted when it should have been rejected.

Figure 5

A Geometric Interpretation of α and β Error

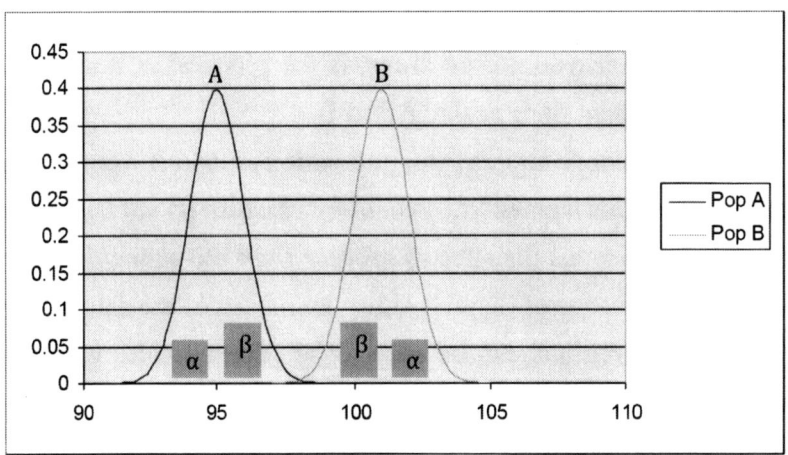

Chapter Two

The foregoing discussion should give you enough information about both hypotheses and t tests to feel secure once you've practiced on some basic data.

My suggestion is for you to make up some paired data sets having various separations between their means and use Excel to run the t test on them.

Always run the F test first to determine whether the variances of the sets can be judged equal and then assign the appropriate number, either 2 or 3 in the final field of the Function Arguments Dialog Box, as per the guidelines in Appendix E.

> Speed is the bottom line of horse racing and racehorse breeding.
> ~ Alexander Mackay-Smith

Chapter Three

Introduction

Experienced handicappers such as Ainslie (211), Beyer (11) and Davidowitz (106) express various attitudes toward speed. In the sense that speed must be judiciously distributed throughout a race and not randomly applied, they are correct.

Some horses run more efficiently when they lead from start to finish—witness Man o' War—others do better staying awhile near the middle of the field, while others lag far behind at first, appearing, then, to make fantastic moves around the field when all seems lost—à la Secretariat's general mode of running.

However, the bottom line is *always* that the horse posting the shortest time—and, thus, the *fastest average speed*—when he crosses the finish line will, by definition, be the winner. There can be no other outcome.

Since only one interpretation and formula exist for calculating average speed, namely, that the distance run equals the average speed multiplied by the time required to run that given distance, there is no other non-contradictory conclusion on this topic.

The Argument about Era Differences

This is as good a time as any for discussing speed as it pertains to the debate over era differences. The era question has been addressed at length in my previous books, GHA and BG. (Justice 107; 122) A slightly different perspective is now offered.

Many advocates of Man o' War as the greatest runner of the past century will absolutely brook no considerations that another horse may have been an equally gifted runner.

Some of these zealots even forbid comparisons by simply stating that one cannot compare horses from different eras. Some even pronounce it *axiomatic* that the comparisons cannot be made, thus venturing into the realm of mathematics where they definitely have no business.

However the admonitions are framed, the dogmatists never define what constitutes an era in horseracing, let alone an era difference. Nonetheless, they keep repeating the same mantra-like utterances.

While it is historically accurate that Man o' War ran fifty-three years before Secretariat—in the years 1919 and 1920 of the past century, a fact is now presented which, in and of itself, casts a dark shadow over such arguments.

Imagine yourself living in the year 1721 in the town of Newmarket, Suffolk County, England.

While ambling around the village market one day you hear talk of an upcoming horse race. There will be three contestants whose names are Devonshire (Flying) Childers, his half-brother, Almanzor, and a brilliant mare owned by the Duke of Rutland named Brown Betty.

The three will race on Newmarket's Round Course. The distance will be the interesting span of 3 miles 6 furlongs and 93 yards, which simplifies to exactly 3.8 miles.

Chapter Three

Flying Childers' impost will be set at 9 stone 2 pounds, which equates to 128 pounds. Almanzor and Brown Betty will each bear imposts of 8 stone 2 pounds, or 114 pounds. (Mackay-Smith 11)

You and a healthy crowd attend the event and witness what would be considered remarkable could any of you peer 291 years into the future.

For this race will be timed, accurate to about 3 seconds in that period (Wikipedia 1), and all posterity can thereby note that Flying Childers, then age 6, completed the 3.8 miles in 6 minutes 40 seconds.

For reasons unknown even to you, you passed on notes of this event which generations of your descendants safeguarded. One day in 2013 an investigator privy to these notes decided to compare Flying Childers' speed with that of "modern" Thoroughbreds.

The American Racing Manual, 2012 edition (ARM 911), was consulted on both World and North American Dirt Records. The results startled him.

He found that the 3-mile world record, set in 1941 at Agua Caliente, Mexico was 5:15. The 4-mile world record, established at Churchill Downs in 2012 was 7:10.80. Thus, Flying Childers' time of 6:40 ± 3 seconds nearly three centuries earlier was at least near current world-record time. The horses setting the modern records carried only 113 and 119 pounds, respectively.

The investigator then compared Flying Childers' time to the current North American Dirt records with similar results. The second set of records showed 3.625 miles, or 29 furlongs, in 6:49.60—slower than Flying Childers—and 4 miles in 7:10.80. The respective imposts were 120 and 119 pounds.

41

To fine-tune Flying Childers' time in comparison to these records, he took another step. He performed linear regression analysis on the North American Dirt records for five furlongs through 4 miles. Figure 6 shows the results.

Figure 6

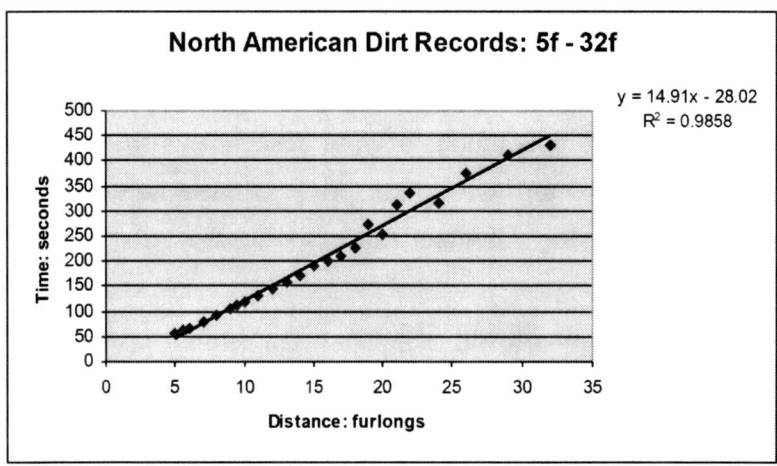

The five-furlong lower limit was used because it included the shortest distance Man o' War ran as a juvenile. It, therefore, provided a convenient boundary point.

To apply the equation, a distance, x, in furlong units is chosen. It is then multiplied by 14.91 and 28.02 is subtracted. The result is the predicted time (seconds) for the given distance—so long as it is between five and 32 furlongs.

The analysis generates an equation that fits its 24 data values with remarkable accuracy, considering that it is based on 26

Chapter Three

different horses (two ties) running 17 different tracks at ages 3 thru 7 (average age = 4.65 years) while carrying an average impost of 119.58 pounds and spanning 71 years. The equation has a coefficient of determination of 0.9858.

It means that it accurately predicts 98.5 percent of the changes in time from changes in the distance run.

When this equation is used to predict the time a modern horse would run the non-standard distance of 3.8 miles—30.4 furlongs —run by Flying Childers in 1721, the result is 425.24 seconds. This means that 291 years *earlier* Flying Childers ran the equivalent distance to modern horses in 25.24 seconds less.

A further calculation shows that, at Childers' average speed of 50.16 feet per second, he traveled slightly more than 1,266 feet in 25.24 seconds. This further means that he would have beaten the modern record-holding horse by about 158 lengths had they run a match race over 3.8 miles.

Learning this undoubtedly made the investigator smile.

Although the preceding interlude into a previous century was imaginary, the data stated are real. It is on record that Devonshire (Flying) Childers ran the cited race in the cited time.

Pause and reflect now on the implications of this lone race, because they touch the major suppositions and conclusions of this book directly.

Evidence against the Era Comparison Controversy

If Man o' War ran in a different era from horses of the latter half of the past century, since he and Secretariat were separated by 53 years, what kind of era gap do you suppose 291 years represents? Yes, that is actually a rhetorical question!

The latter time difference would most certainly reflect, if anything substantial existed to be reflected, an era difference—and yet a horse running a track that, with near certainty, was inferior to "modern" tracks in every manner established a time that easily bests the modern times for the same distance.

Even if one argues that the likely 3-second error of the timing device aided him, Flying Childers still holds a 23-second advantage over current record-setting horses.

Furthermore, Flying Childers carried 128 pounds. Sotemia, the current horse holding both the North American and World Dirt records at 4 miles carried 119 pounds.

And, although Flying Childers was six years old when he ran his cited race, Sotemia was 5. Therefore, it cannot reasonably be argued that Flying Childers was significantly more mature since age 5 is considered full maturity. Besides, arguing from that perspective is stretching the imagination.

Regarding weight-carrying ability, only Dr. Fager, under a 134-lb. impost, established a world record anywhere nearly as impressive as that of Flying Childers. That record was set at Arlington Park in August of 1968. Dr. Fager ran one mile in 92.2 s. The record has been tied, but by a horse named Najran under only a 113-lb. impost.

Another interesting calculation can be made for comparison purposes. The standard formula for kinetic energy from basic physics can be used to calculate how the average energy maintained by Dr. Fager and Flying Childers compare for their respective races.

Chapter Three

Since changes in kinetic energy equate to work done against mainly retardant track forces, using the Work-Energy Theorem (Sears 120) gives a good idea of how the two horses overcame the track resistance of the surfaces they ran.

Applying the Work-Energy Theorem

The Work-Energy Theorem states that the *change* in kinetic energy from the beginning to the end of some process equals the *total work* done by the resultant force during the process. In other words, work *is* change in kinetic energy.

We now arbitrarily adopt a convention requiring the supposition that both Flying Childers and Dr. Fager were capable of running two furlongs in current world-record time. This assumption is not far from correct and it is relative anyway. However, it provides a common baseline for comparing both horses.

The current world-record for two furlongs, the shortest distance recognized for world-record purposes in the American Racing Manual, is 0:20.57. It was set by two-year-old Winning Brew at Penn National track on May 14, 2008. (ARM 911)

If Flying Childers and Dr. Fager both ran at the implied average speed for two furlongs, 64.17 ft/s, their respective average kinetic energies—the energies they imparted to their jockeys plus tack - for those two furlongs *would be* 8,184.43 units and 8,568.07 units, respectively. These *hypothetical* kinetic energies for two furlongs provide the starting baseline for the Work-Energy Theorem.

The *actual working* kinetic energy is calculated using the formula 1/2 • mass • speed (or velocity) squared. In symbols it is: $½ • m • v^2$.

45

In the formula, the mass (impost) carried by each horse is simply the equivalent weight in pounds divided by 32.2, the average acceleration of earth's gravity at sea level. Units (ft/s^2) for acceleration may be omitted if consistency is maintained.

Flying Childers, carrying 128 lb., therefore carried a mass having 3.98 units at an average squared speed of 2,516 units for 3.8 miles in a total of 400 s.

Again, for this calculation exact units are omitted because they can be confusing to the uninitiated reader. Kinetic energy units were fully explained in Chapter 1. The present conclusion remains unaffected by temporarily dropping units.

Multiplying 1/2 · 3.98 · 2,516 gives Flying Childers' average generated kinetic energy of 5006.84 units for his 3.8-mile run.

The corresponding mass and squared-velocity units for Dr. Fager are: 4.16 and 3,279.49, respectively. Using the kinetic energy formula on these values gives 6,821.34 units for Dr. Fager's average kinetic energy for one mile.

The *total changes* in average kinetic energy between the start (2-furlong reference) and finish of each horse's races are: 8,184.43 - 5,006.90 = 3,177.59 and 8,568.07 - 6,821.30 = 1,746.73 for Flying Childers and Dr. Fager, respectively.

Dr. Fager's energy was maintained for 6 f, the distance from the two-furlong reference point to his one-mile finish. Flying Childers maintained his kinetic energy over 28.4 f, the distance from the two-furlong reference point to his finish, which was 28.4 f further.

Dividing each horse's total kinetic energy change by the respective number of furlongs run, one obtains -111.89 units per furlong for Flying Childers and -291.12 units per furlong for Dr. Fager.

Chapter Three

It is, therefore, arguable that Flying Childers maintained a *lower kinetic energy loss per unit distance* than did Dr. Fager, especially since his "track" at Newmarket, England was undoubtedly less conducive to efficient running than was Arlington Park where Dr. Fager ran some 241 years later in 1968.

Thoroughbred Improvements over Three Centuries

Has the Thoroughbred, as an overall breed, *really improved significantly* since Spanker, the first true Thoroughbred, was foaled about 1670 at Helmsley Stud in northern England? The date of Spanker's foaling is not explicitly recorded in the General Stud Book, but it was near the middle reign of Charles II (1660 – 1685). (MacKay-Smith 3)

It is easily shown, as was done in GHA, that a moderately strong correlation exists between annual foal crop size changes throughout the twentieth century and the corresponding lowering of average racing times for several major races.

This trend does not mean, however, that each successive foal crop generates at least one horse superior to those in all preceding generations. It more probably means that the top few percent of the best runners in any given crop are simply greater in number. This can easily account for the aforementioned trend.

Although by no means proof of the foregoing concept, the time that Flying Childers clocked for 3.8 miles in 1721 presents a convincing argument that he was at least an equivalent runner to any modern Thoroughbred.

This issue will be expanded when the topic of "absolute" track speed is discussed in conjunction with Man o' War's and Secretariat's running times for equivalent distances in Chapters 6 and 7.

As previously noted Hollingsworth estimates that only 2.5% of horses in a given foal crop ever win a single stakes race. (35) Ours doubles this figure to 5% (62).

Whatever the exact number, it represents a low percentage.

Since Man o' War's 1917 (his foaling year) foal crop contained 1,680 foals, it was the second smallest crop between 1897, when the Jockey Club registration process began, and the present. The only smaller crop was that from 1919 which lists 1,665 foals. (ARM 719)

Taking 2.5 percent of the 1917 crop gives a possible 42 horses, colts and fillies combined, that could become single stakes winners.

Thus, each time Man o' War raced, it was possible, though not strictly provable, that one or more of the remaining 41 candidate horses from his foal crop would face him. These challengers included Upset, the only horse to beat him, and John P. Grier, the only horse that probably pushed him to his speed limits in the 1920 Dwyer Stakes.

Secretariat's foal crop from 1970, his foaling year, numbered 24,361. Applying the 2.5 percent rule-of-thumb indicates that 609 potential single stakes winners were within that crop.

Thus, each time Secretariat raced, potentially any of 608 other colts or fillies were likely opponents.

Among these potential opponents were: the excellent bay colt, Sham, who was thrice overshadowed by the brilliant chestnut from The Meadow in the Derby, the Preakness and the Belmont, Herbull, who beat Secretariat in his maiden race because he was nearly knocked down at the starting gate, Stop the Music who won the Champagne Stakes via Secretariat's disqualification, Angle Light who wrested the Wood Memorial from him because of an untreated mouth abscess, Onion who took the Whitney from a feverish Secretariat and Prove Out who triumphed in the Woodward, ridden by Jorge Velasquez, over a tired Secretariat. (Wolfe 162)

The Distribution of Talent among Foals

If one uses Excel's NORMINV function and enters the probability value 0.025 in the first array window, a mean value of 0 in the second array window and the value 1 in the final window, the answer, -1.96, rounded to two decimal places, is obtained.

This is the number of standard deviations below the mean of any standard normal distribution assumed to show the apportionment of Thoroughbred running ability.

If Hollingsworth's 2.5% estimate is reasonably accurate, it implies that in the 1917 crop 42 horses, including Man o' War himself, were at or below the -1.96 standard deviation level of their distribution with respect to racing ability.

This means that the remaining 1638 horses had little or no chance of winning one stakes race among them.

For Secretariat's 1970 crop of 24,361 foals, 609 of them fell at or below the magic -1.96 standard deviation level of talent. That leaves 97.5 percent, or 23,752 horses having slight chance of winning a single stakes race. Refer to Appendix G for a brief description of how to generate the standard normal curve using Excel.

Such numbers are stark reminders of the vagaries of chance in the world of Thoroughbred racing.

It is appropriate now to investigate further the implications of the supposedly innocent assertion alluded to above—that racing talent throughout a Thoroughbred foal crop is normally distributed.

The concept of track variant will be analyzed as it relates to this distribution model to determine what it discloses regarding relative talent of horses from different eras.

Track Variant as Estimator of Relative Ability within Foal Crop

Track Variant is one of the numbers included by the Daily Racing Form in its past performance sheets.

It is coupled with Speed Rating to produce an entry having the form abc-de, where 'a' through 'e' can be any of the digits 0 through 9. Typical combinations might be, for example, 98-07 or 101-14, where the number before the hyphen is the Speed Rating and the following number is the Track Variant.

Speed Ratings compare a horse's finishing time to the *fastest time* at the given track for the same distance and track condition *in the last three years*. The fastest time is given a 100 rating. A 100 rating is often called 'par.' One point is then *deducted* from 100 for *each one-fifth* of one second (0.20 seconds) by which a horse fails to match that time. Therefore, if the winner of a race ties the best time, his Speed Rating is 100. If he is three-fifths of a second slower, 0.60 seconds, his rating is 97.

Track Variants attempt to complement the Speed Ratings. To quote Daily Racing Form: Track Variant takes into consideration

Chapter Three

all races run on a particular day under the same conditions of distance and track surface. If the definition stopped there, all would be clear. However, the same paragraph adds: *If the average Speed Rating of winners sprinting on the main track is 86, the Track Variant is 14 (par of 100 minus 86).* (28)

The generic word 'sprinting,' unqualified, is what makes the interpretation ambiguous. If DRF simply added ". . . the Track Variant for sprints *only* is 14," all would be well.

However, a typical race card for a day can easily have eight or nine races of mixed distances, not just sprints. Sprints, by definition, do not include any distance beyond seven furlongs.

Since the typical past performance chart tells nothing about the number of sprint races, nor their range of distances, for the day from which the average Speed Rating was calculated, then all one knows is that the average of the Speed Ratings is the Track Variant value. It tells nothing about the variations of the track *for a specific race* on the given day.

Even highly knowledgeable handicappers admit doubts about the definition and use of the Track Variant and its meaning or application. (Ainslie 216)

However, in this book, it is assumed that the Track Variant gives a stable enough reference to use for its purposes—giving time credit to Man o' War for possibly slower tracks.

To show the likely fluctuations that accompany the Track Variant, 10 random numbers were generated using Excel's RNG. A mean value of 91 and a standard deviation of 3 were *assigned* to this simulation. These parameter values were chosen to assure results that could be realistic Speed Ratings for a given day at some imaginary track.

If the 10 random numbers are selected from a normal distribution having the stated parameters, one could expect values ranging from 82 to 100, given a standard deviation of 3.

The values actually generated were: 93, 86, 93, 93, 94, 96, 88, 92, 93 and 91. The mean value of these 10 hypothetical Speed Ratings is 92. Therefore, if a certain track produced these results for 10 *sprint races* on a given day, the Track Variant for that day for sprint races *only* would be 100 − 92 = 8.

However, the low value in this group is 86 and the high value is 96. There is, therefore, a range or data spread of 10 points within this set of values.

And so it is well and good to say that the Track Variant for this example is 8, but that by itself obviously tells nothing about the actual pattern of sprint races that day

Last, a low Track Variant can mean faster track surfaces *or* higher quality runners for the day—or both. For a Track Variant of 00 to be possible, the winner of each race of the same category (sprint or route) for the day would have to tie the track record. This would give him a Speed Rating of 100. That is the obvious way the average Speed Rating for the day could equal 100 and, therefore, make the Track Variant equal 00, or 100 − 100.

Track Variants for Man o' War's Six-Furlong Races

Man o' War ran six six-furlong races as a juvenile in 1919 between July 5 and September 13. The track distribution for these races was: Aqueduct - 1, Saratoga - 4 and Belmont - 1. His rest period between these races averaged 14 days. (Daily Racing Form 28)

For these six races, the combined Speed Rating/Track Variant number pairs are, in the order just given for the chronological racing dates: 90-12, 90-13, 95-09, 92-08, 87-14 and 85-21.

Chapter Three

We can use both numbers from each pair to determine Man o' War's z-score for each of these races. This basic procedure is done for Man o' War and Secretariat in Chapter Eight.

The z-score *gives a direct relative ranking* of Man o' War with respect to his peers. This ranking, in turn, provides at least an *approximate* general test of Hollingsworth's previously noted statement regarding 2.5% of foal crops and the winning chances for stakes races.

Looking ahead, his average z-score for these six races was -1.023. That value immediately places Man o' War slightly more than one standard deviation *below* his peer average for the six races. Remember that 'below' average for z-scores is always better.

When Excel's NORMDIST function is used to find the cumulative percent of area beneath the standard normal curve and less than the 'x' value -1.023, the result 0.1532, rounded to four decimal places.

This indicates that about 15 percent of Man o' War's peer group could theoretically be expected to run faster six-furlong races than he. This at least gives a rough estimate of overall peer quality within his 1917 foal crop, but it cannot pinpoint what percent of the foals within that crop could win at least one stakes race.

If I wanted him to jog, he wanted to gallop. No matter what I wanted, he wanted to go faster.
~ Louis C. Feustel

Chapter Four

Man o' War

This chapter and the next are biographies of Man o' War and Secretariat, respectively. The main reason that a kind of judicious brevity influences them is that, as of this writing, a leading internet book vendor lists 253 operative titles for Man o' War and 185 for Secretariat.

There is little reason to amplify subjects which already saturate the market from differing viewpoints.

Man o' War was foaled at Nursery Stud, near Lexington, Kentucky, just minutes before midnight on March 29, 1917. He was the product of a cover arranged by his breeder and prominent New York banker and financier, August Belmont, Jr., between the stallion, Fair Play, and the dam, Mahuba.

Nursery Stud was first established in 1867 at Babylon on Long Island, New York by August Belmont, Sr. He later moved it in 1885 to its location about five miles from Lexington. His son, August Belmont, Jr. took over its lease in 1897, some 20 years before Man o' War was foaled. (Hollingsworth 119)

Man o' War & Secretariat

Man o' War's tail-male lineage traces to Matchem, who was a 1748 grandson of the Godolphin Barb. (Ainslie 79)

The Godolphin Barb (1724-1753) is more often referred to as the Godolphin Arabian, but he was not Arabian. His importer, and breeder-owner, Edward Coke, referred to him as Arabian in order to charge higher stud fees— apparently something not attended to by most interested patrons. (Mackay-Smith 132)

Such considerations aside, the Godolphin Arabian, the Darley Arabian and the Byerley Turk are basically considered the genetic bedrock of the Thoroughbred horse for England and, eventually, for North America. (Ainslie 78)

It should be noted, however, that another stallion, Curwen's Bay Barb, is responsible for a somewhat higher percent of genes in the current Thoroughbred population than is the Byerley Turk. The respective percentages are: 5.6 and 4.8. (Cunningham 94)

With respect to tail-male lineage, Man o' War generally differed from the equally gifted Secretariat whose equivalent ancestry traces to Eclipse, (1764-1806) a great-great grandson of the Darley Arabian. As Ainslie notes (79), however, typical Eclipse descendants nearly always include numerous connections to Matchem (1748-1781) and Herod (1758-1780).

Matchem descends from the Godolphin Arabian, while Herod was a great-great grandson of both the Byerley Turk and Curwen's Bay Barb on his sire's side (Tartar) and of the Darley Arabian on his dam's (Cypron) side. (Wikipedia)

To make the most emphatic case for a common lineage between Man o' War and Secretariat, consider that Secretariat

Chapter Four

is directly related to Fair Play, Man o' War's sire, through Miss Disco, the Dam of Secretariat's Sire, Bold Ruler. (Mackay-Smith 171) More will be said on this point later. This fact makes it unequivocal, without further ado, that Man o' War and Secretariat shared common genes.

Apparently Mr. Belmont originally planned on keeping Man o' War and 21 other yearlings for his own racing purposes but decided it was best to sell them when he felt compelled to join the Army Quartermaster Corps as major and serve in Spain during World War I. He was also undergoing some financial strain from a troubled Cape Cod Canal project in which he had invested heavily and which was doing poorly. (Hollingsworth 115)

Since Mr. Belmont was probably what might now be termed a "Type-A" personality, he judged that he could not attend to the business of overseeing the development of these horses properly and thus decided to sell them.

They were subsequently shipped to Saratoga, New York for and auction by Powers-Hunter Company set for Saturday, August 17, 1918. The group of 22 sold for $52,250 with the "baby" Man o' War, though he seemed somewhat anxious and rambunctious, commanding the third highest bid of $5000. (Ours 31) Hollingsworth gives the group sales figure as $51,450. (Hollingsworth 120)

Therefore, on the day and near the site of the 49th running of the Travers Stakes, the glistening chestnut yearling— nearly 17 months old— became the property of Samuel Doyle Riddle, a prominent fabric merchant and connoisseur of fine equines from the town of Glen Riddle, Pennsylvania.

Now under the immanent guidance of trainer Louis C. Feustel and jockey John Loftus —both future racing Hall of Fame inductees— Man o' War was about 10 months removed from that special touch of fame reserved for a slim fraction of all Thoroughbred runners.

His striking name, selected by Mrs. Eleanor Robson Belmont to honor her husband's war participation, could not help but add to the mystique of the near unearthly powers he seemed to unleash on the track.

But all that was future. It would be nearly three months before the war subsided and peace treaties were signed. It would be nearly eight months later, on June 6, 1919, before the two-year-old Man o' War first appeared on a race track.

Sandwiched between the present doldrums and the future glory was the relatively painstaking process of "breaking" the colt— first to allow a bridle and saddle to touch him for more than a few seconds and, only then, to suffer the burden of a man astride his back urging him to do what he actually wanted to do more of than the man wished—run!

At first he disliked both experiences, and he was not reticent in letting his human managers know of his preferences. (Farley 109)

Somehow it all coalesced, we humans judge with hindsight, as it was meant to be. A still impatient but mainly cooperative

Chapter Four

Man o' War stood behind a net barrier at a place called Belmont Park on Long Island in New York State.

The day was Friday, June 6, 1919. Six other fledgling runners fidgeted in various modes of disarray alongside him. They were supposedly being poised to begin an ungainly-sounding adventure termed a 'maiden special weights' five-furlong race.

And then, in a peculiar way, it all seemed nearly anticlimactic and even foreordained. After the net shot upward, it took but fifty-nine seconds for onlookers of racing discernment to realize they had watched an extraordinary chestnut two-year-old grab the bit, swoop along the track and cross the finish line six lengths ahead of his nearest rival, Retrieve. It was no contest. The daily racing form, even more laconic, noted simply 'Easily.'

His legend had, in fact, begun that day on Long Island's soil.

From June 6, 1919 until October 12, 1920 the name 'Man o' War' held the hearts and minds of Thoroughbred lovers as perhaps no other— at least in American track annals— horse had ever done.

In 21 starts, apportioned 10 to the juvenile and 11 to the sophomore racing years, he won 20. His lone defeat was somewhat ironic in that the synonym for an unexpected loss— an 'upset'— was also the name of the horse who accomplished it.

But 16 months and 21 total races pass quickly. Then they are over for the most part. In Man o' War's case, since he was the century's superstar of racing, his career at stud was poised to begin.

Compared, however, to his racing accomplishments his procreative powers almost had to take second place.

He did sire some outstanding competitors. Among these were War Admiral, probably the most famous, and the grandson, Seabiscuit, who was a close second.

Three incidents in his racing career are more noteworthy than his breeding success. These are now discussed before a truly magic time in his life, his final 16 years with a generally unheralded but unique and inspiring man, Will Harbut, conclude his story.

Juvenile Year Anomaly: The Sanford Stakes

By August 13, 1919 Man o' War was an unbeaten veteran of six races. The public and sports writers were already fairly convinced that he was probably the greatest Thoroughbred ever to have run on American, or any other, soil.

He constantly gave his trainer and jockey the feeling that he always was ready and able to give more than they ever asked. Indeed, he seemed bent on showing them that he was possessed of a flame so intense that he was frustrated because it was kept in check.

The racing public had much this same conception because of the nearly disdainful manner in which he seemed to dispatch opponents— routinely eased, for instance, through the final 16th of his six-furlong races and displaying a certain 'haughtiness' in the way he carried himself and ran.

Just eleven days previous, on Aug 2, Man o' War ran a six-furlong sprint at Saratoga in 1:12.40. His speed rating for that run was 90— meaning that it was two full seconds slower

Chapter Four

than the existing track record, with each point of speed rating representing one-fifth of a second.

However, his comment line shows that he was eased the final 16th mile. In addition, he also carried 130 pounds. Had he been allowed to run, despite that impost, he possibly would have owned a new track record— a conclusion readily adopted by his growing legion of admirers.

In the August 2 event, the U.S. Hotel Stakes, he handily beat a horse named Upset by 2 lengths while carrying 130 pounds to 115 for Upset. It was then another length from Upset back to the show horse, Homely, carrying 112 pounds and still another length back to the first of the remaining seven field horses. (Daily Racing Form 28)

Given these facts, it was fairly assumed that the upcoming race, the Sanford Stakes, would present little drama or novelty for the glistening and imperious chestnut. It would simply provide another opportunity for Man o' War's supporters to tout his superiority, for Upset, already beaten by Man o' War, was entered.

Whatever the real truth, the presumed inevitable didn't occur. Opinions were divided basically between a poorly managed start and either inept horsemanship or deliberate throwing of the race by John Loftus.

That John Loftus, known since his early racing days as impeccably honest, would throw a race, is nearly unimaginable through hindsight.

Put simply, Man o' War lost his first and only race to Upset. John Loftus was not asked to be his jockey for the next year. Clarence Kummer would be up.

It so happened that the starter that fateful day, Charles H. Pettingill, was a substitute for Marshall (Mars) Cassidy who was said to have contracted tonsillitis the night before and could not fulfill his normal duties.

Mr. Pettingill had once been a prominent starter but was nearing retirement and not in prime form. He now served mainly as a placing judge who sat in the stands at the finish and noted which horse first crossed the finish line.

All such information may be correct, but it does not actually state whether Man o' War was distinctly disadvantaged at the start, as many claimed. One general interpretation was that he was actually facing away from the barrier when the race began, thus having to turn in the correct direction first and then chase after the field. (Ours 102)

If that were true, the colt was distinctly not used to this situation. John Loftus routinely spent many hours training him to actually charge "under and through" the barrier by lifting it with his head as an assist to its normal spring-activated movement.

On more than one occasion, Marshall Cassidy had noted after a race that he nearly recalled the horses due mainly to Man o' War's quick start. However, he never did.

Undoubtedly the most sober account of the actual Sanford start was by Louis Feustel, Man o' War's trainer. According to Ours, Feustel simply termed it a "rather staggering" start when later interviewed on the issue. (Ours 102)

A Closer Look: the Sanford and Four other Six-Furlong Races

It is informative to review the past performance data on the Sanford with respect to four other six-furlong races run by Man o' War between July 5, 1919 and August 30, 1919.

Chapter Four

Man o' War actually ran his last six-furlong race, and the final race of his juvenile campaign, on September 13, 1919 at Belmont Park. However, only the finish time is given by Daily Racing Form for that final race, and a single number cannot be used, as is the data from the other six-furlong races, to form a separate linear regression for each race.

Such single-race regressions can show how time progressed as Man o' War passed each separate furlong point, not just the regular call points, throughout each race. From those progressions, one can draw some important conclusions of just how important a "flying" start may or may not have been to Man o' War.

Data List 1 contains the information needed for this somewhat abbreviated analysis. The interpretation for the list's notation is: Post = post position for Man o' War; Barrier Exit = Man o' War's field position just after leaving the barrier; call notations like **1**:1.1, read left-to-right, mean that at call position number 1 (bold font), Man o' War was first by one length, thus **1**:1.1; at call position 2 he was first by one length, thus **2**:1.1, etc. Other headings should be self-explanatory. In line four, 'hd' means 'by a head.' The third line of the list, date shaded, refers to the Sanford Stakes. In that race the call times refer to the leading horse, Upset, vice Man o' War.

Data List 1

Date	Track	Condition	Call 1	Call 2	Finish	Post	Barrier Exit	Call Positions
5Jly19	Aqu	fast	:23.6	:47.4	1:13.0	2	2	1:1.1, 2:1.1, 3: 1.1.5, 4:1.1
2Aug19	Sar	fast	23.0	:47.2	1:12.4	8	1	1:1.3, 2:1.3, 3:1.4, 4:1.2
13Aug19	Sar	fast	23.4	:47.1	1:11.3	6	5	1:4.1.5, 2:3.2, 3:3.1.5, 4:2.0.5
23Aug19	Sar	fast	22.6	:46.4	1:12.0	2	3	1:hd, 2:1.3, 3:1.3, 4:1.1
30Aug19	Sar	sl	:23.3	:47.3	1.13.0	3	4	1:2.2, 2:2.1, 3:1.5, 4:1.4

Several list items immediately stand out. First, Man o' War actually had one less favorable post position than in the Sanford Stakes, that being in the 2Aug19 U.S. Hotel Stakes. He broke first from the barrier in the latter race, the only time in the five races when he did so.

The call positions for the Sanford show, in accordance with written accounts, that he was gaining on Upset throughout the race but that he lost by one-half length.

It is also interesting that Man o' War's finishing time for this race can be calculated— it is not given explicitly in the past performance chart, since he did not win — as 71.27 s versus 71.20 s for Upset. This is actually the fastest juvenile time for six furlongs that Man o' War ran. He obviously did so because he was behind and trying to catch the leader. In horse racing jargon he was "set down."

When a linear regression is performed for each of these races, using the call point times (separate call point times must first be determined for Man o' War's Sanford since he was not leading at these junctures and they are not explicitly given by Daily Racing Form), one can see from the regression coefficients that the Sanford Stakes was, despite his loss, actually his *best six-furlong effort*.

Chapter Four

Linear regression analysis shows that his coefficient of determination, COD, for all five races was well above 99.9. This means that the separate regression lines predict his changes in running time for the various call points with nearly 100 percent accuracy. In fact, the average COD for the five races is 0.9997 rounded to four decimal places.

This number represents extraordinarily well-fitted regression lines. Such lines may be used with confidence for further interpretations of Man o' War's running prowess in those races.

In that regard, the Excel TREND function was used to generate further information of his exact placement throughout each race. The original data included only three call points, the quarter and half mile, plus the finish. The TREND function allows one to choose arbitrary distances within any race to check on a horse's placement relative to the starting and ending points of that race.

Exact single-furlong locations from 1 through 6 were chosen in these cases. In fractional miles, those distances are: 0.125, 0.25, 0.375, 0.50, 0.625 and 0.75, the finish.

Data List 2 showing Man o' War's *projected times* (in seconds) for, these distances within each race follows. The list values are from the aforementioned linear regressions. They differ slightly, therefore, from the actual times in the DRF past performance record.

In the list, all distances are in fractional miles and corresponding times are in seconds.

Data List 2

Distance	5Jul19 Time	2Aug19 Time	13Aug19 Time	23Aug19 Time	30Aug19 Time
0.125*	10.95	10.48	11.36	9.95	10.57
0.25	23.30	22.83	23.33	22.30	22.99
0.375	35.65	35.18	35.29	34.65	35.43
0.50	48.00	47.53	47.26	47.00	47.86
0.625	60.35	59.88	59.22	59.35	60.29
0.75	72.70	72.23	71.19	71.70	72.72

* Time inaccuracies on this line are due to the distance being beyond the regression limits.

It is immediately evident from the list that Man o' War's *projected* losing time for the Sanford Stakes, 71.19 s, ironically represents his best juvenile effort in a six-furlong sprint. His next best time for the distance occurred ten days later in the Grand Union Hotel Stakes in which he beat Upset by one length.

Even though Upset carried 125 pounds to Man o' War's 130 and the comment line has Man o' War being eased the final 16th mile, it was still a respectable performance by Upset, a horse not considered a major champion by present standards.

The winning time difference between the two consecutive races featuring Upset was 0.80 s. That time is the *actual race time difference* as opposed to the time differences in Data List 2 which were projected times using Excel's TREND function.

Both races were run at Saratoga, and the track was listed as fast for each. Upset won the Sanford Stakes with an earned speed rating of 95. Ten days later Man o' War beat Upset by a length at the same distance and earned the lesser speed rating, 92.

What all this shows is debatable, other than the "luck" that horseplayers often like to quote. It is the kind of information that

Chapter Four

speed handicappers such as Beyer (11) and Davidowitz (107) generally fault. Ainslie clearly explains such results (212)

Superficially, one might expect a more uniform effort from both horses within a ten-day period on the same track and under reasonably equivalent surface and impost conditions. Truisms, however, live briefly in Thoroughbred racing.

Perhaps all that can really be deduced from such analyses is that the multiple, independent variables affecting each race make it simply too difficult to offer sweeping generalizations about outcomes with any real degree of accuracy.

An interesting and informative pattern can, however, be found from the list by taking successive correlations between the times for each intermediate furlong distance and the finish time for all five races.

The respective furlong-finish correlation pairs are: -0.1846; 0.1255; 0.5530; 0.8636; and 0.9803. Recall that the equivalent fractional-mile distances are 0.125, 0.25, 0.375, 0.50, and 0.625, respectively.

The correlations grow progressively larger *and* progressively more positive as the race develops. The negative, small value for the first correlation may indicate that scant relationship holds between the first or second furlong time and the finishing time—at least in Man o' War's case.

By the third furlong, however, the correlation 0.5315 "explains" over 28 percent of the finishing time value. That is the halfway point. One would expect that "pace strategy" would acquire meaning by mid-race.

Recall that the square of the correlation coefficient, the COD, is used for judging explanatory value, in terms of prediction accuracy, and not the correlation magnitude itself.

This is interesting from another standpoint. Namely, Ours (101) states that Man o' War's jockey, John Loftus, was told not to let Man o' War go until the third furlong. Although his owner and trainer undoubtedly did not study statistics, their combined racing experience told them that this was generally a critical point for the distance.

By the fourth furlong some 74 percent of the finishing time is determined by the time at that point, and by furlong five nearly 96 percent of the outcome is settled.

These figures seem to be consonant with what common sense might say about such races.

In closing on Upset during the Sanford, Man o' War ran a final furlong in 12.05 s at an average speed of 54.78 ft/s—his fastest final furlong in the entire set of six-furlong races.

An important fact emerges from this data when Man o' War's times for the quarter mile are compared with the fastest times for the year 2011 at the same distance. The comparison is not as inappropriate as it may first sound.

The American Racing Manual (914) lists 11 such times set at 11 different dirt tracks. The times range from 19.30 to 22.47 seconds. Their average is 21.50 s and their standard deviation is 0.85 s.

Recall that these are quarter-horse times, times set by horses specifically bred for all-out speed at this short distance.

The times at two furlongs for Man o' War are entirely respectable compared to those of the quarter horses. In fact, his

Chapter Four

fastest two-furlong time, 22.60 s, is 1.30 standard deviations above the mean of the quarter horse times. This is well within the expected plus-three-sigma level for a normal distribution.

Since his impost in all five races was 130 pounds, and the North American dirt record of 0:20.57 s for two furlongs was set on May 14, 2008 by Winning Brew carrying 119 pounds, Man o' War's efforts at the start of his six-furlong races were impressive.

Another issue is, however, raised by Man o' War's excellent times at the two-furlong point. That is, were the tracks at Aqueduct and Saratoga when Man o' War raced slow relative to the ten tracks cited in 2011 as having the fastest two-furlong times?

It is a question to be kept in mind and investigated later. It does seem that Man o' War did not suffer an undue disadvantage from much slower tracks since he maintained a reasonable proximity to times for the same distance some 90 years later.

First Anomaly of the Sophomore Year

Equally intriguing as was the Sanford Stakes analysis is the question of why Man o' War did not enter the 1920 Kentucky Derby.

Ours (141) gives the general interpretation that Samuel Riddle did not believe that his colt should shoulder the 126-pound impost so early in his sophomore year and for so long a distance. Riddle thought this would strain Man o' War too much.

Thus it happens that the great Man o' War—bred, born and raised in Kentucky, never raced there. Consequently, although it would have been "after-the-fact" as it was for the cases of Sir

Barton and Gallant Fox, he never had a chance to win the Triple Crown.

It is nearly certain that he would have accomplished the feat. The eventual winner, Paul Jones, took that first jewel of the Triple Crown—not then yet established—on a muddy track in the time of 2:09.00. None other than Upset placed and On Watch showed.

Man o' War had lost only in the Sanford to Upset. He had beaten On Watch twice as a juvenile, and he defeated Upset by one and one half lengths in the Preakness two weeks later at 9.5 furlongs. His comment line for that win was "speed in reserve."

Skipping the Kentucky Derby does at least seem odd, even though it did not quite hold the same aura as today, since just two weeks later Man o' War carried 126 pounds in the Preakness. We are only considering two races having no impost difference, differing by only one-sixteenth of a mile and separated but two weeks in time. Yet Man o' War ran in, and won, the Preakness.

Ours hints about the reason for Riddle's decision when she notes that New York horsemen still called Kentucky "the West," as though it were still a bit wild. (141). Perhaps a feeling yet lingered within the easterner, Samuel Doyle Riddle, that a 'western' race was not consequential enough for his colt? Whatever the true reason, citizens of the Commonwealth of Kentucky never saw Man o' War race there.

In fact, Man o' War was probably nearing the peak of his sophomore maturation process by then. Ours notes that, seventeen days before the Preakness, horsemen who watched him work, having come off his wintering in Berlin, Maryland, saw that he was three inches taller, that he had thickened noticeably and that he resembled a four-year-old. (145)

Chapter Four

This point also belies comments of those claiming he was undernourished due to the period in which he lived. This issue will not be elaborated upon here. It has already been sufficiently refuted. (Justice 121)

It is almost as assured as anything in Thoroughbred racing that Man o' War would have taken the Kentucky Derby and, thereafter, the remaining two "jewels" of a Triple Crown yet to be fashioned by Cartier. (Drager ix)

In the ten races following the Preakness and also closing his racing career, Man o' War's average speed rating was 109 with a low of 86 in the Stuyvesant Stakes and a high of 134 in the Lawrence Realization Stakes.

Man o' War ran a thirteen-furlong Realization in 2:40.80. This stakes record was never actually broken. It held until Kelso tied it in 1960, and no horse until Kling Kling in 1970 ran it faster. However, Kling's Kling's fairly short-lived record time of 2:27.60 was at its reduced, and still current, distance of twelve furlongs rather than thirteen. (2005 ARM 1347)

This race constitutes what is, to my thinking, another anomaly, and a rather sad one, in the career of this outstanding champion. It deserves a separate discussion.

Second Sophomore Anomaly: The Lawrence Realization Stakes

This race may, in some sense, be considered the definitive moment in the racing career of Man o' War. If the word 'speed' ever applied to a Thoroughbred, it applied to his performance on that September 4, 1920 day at Belmont Park in New York.

Several mitigating factors, however, should be mentioned in this regard.

By all accounts, the Belmont track, generally considered fast in comparison to other venues, was especially suited to running on this day (Ours 208). A healthy crowd was ready to assemble, but past experience indicated that spectators did not take well to so-called walkovers.

In a walkover, the fame and racing prowess of one horse, in this case Man o' War, literally kept other owners from entering horses against him, so concerned were they of being humiliated or of having their horse's racing spirit forever marred.

Within one hour of the appointed race time, the only other horse entered in the race was withdrawn. George D. Widener, owner of Sea Mint, decided that his horse had shown little enough aptitude for racing thus far that it might be imprudent to enter him.

The only other viable contestants, John P. Grier and Donnacona, were likewise not scheduled to run in the Lawrence Realization.

Fortunately, Sarah Jeffords, niece of Samuel D. Riddle, stepped up, front and center, to avoid the distasteful walkover.

She offered her horse, Hoodwink, for the trial. This gesture also assured that the winner would take home a full purse rather than a half purse for only a walkover.

As if facing the great Man o' War was insufficient burden for what was a marginally good horse, Hoodwink had competed rather respectably just twenty-four hours previous in a six-furlong sprint.

Not only was he just a good horse, he was also a tired horse, plain and simple.

However, Sarah Jeffords and Samuel Riddle saw fit to pit Hoodwink against Man o' War. Methinks the outcome was perceived as obvious before the start.

Chapter Four

One must wonder, given any sense of sportsmanship and since the crowd only wanted to see Man o' War break a world record, whether the better part of "gallantry," for lack of a more apt term, should have prevailed and the unpopular walkover undertaken.

As expected, the fiery chestnut who had lost only one race, and that probably due to poor riding management, made a mockery of whatever dignity Hoodwink might have possessed.

Man o' War did set a world record—smashing the previous mark for 13 furlongs by 6.8 seconds, as his speed rating of 134 indicates. One wonders, however, whether that should be applauded under the circumstances.

He also is credited with "winning" by the ludicrous margin of 100 lengths, depending upon the source, over Hoodwink, a horse obviously so far beyond his competitive ability that terms such as "farcical" come to mind as appropriate descriptors of the result.

Nonetheless, Man o' War's time in the 1920 Lawrence Realization became one for posterity. In 1970 the 'Realization' was changed to 12 furlongs. It remained at that distance until 2005, its final year of running—apparently due to dwindling resources and, possibly, interest, when it was shortened from thirteen furlongs.

Thankfully it was not dropped because a later-day equivalent to Hoodwink ran against Secretariat!

Regarding a descriptor such as "tie," it must be noted that Kelso's comment line states "speed in reserve." Astute readers will understand the implications.

Ours (211) gives fairly elaborate accolades to, and breakdowns of, Man o' War's efforts and times by furlong for the entire race: two furlongs in 0:23.60; three furlongs in 0:35.60; the half mile in 0:47.80. Then fifth and sixth furlongs both in 12.60 to place him at the three-quarters pole in 1:13.00.

Pause here to note that Secretariat has held the fastest six-furlong time at Belmont Park since the Belmont Stakes of June 9, 1973. His time was 1:09.80 s.

From there Man o' War clocked successive furlongs, seven through thirteen, in 13.00, 12.60, 12.40, 12.60, 13.00, 12.20, and, last, 12.00. All totaled, his speed legacy at Belmont Park equaled 2:40.80.

It is instructive to note the power of linear regression by using Man o' War's times at the Realization call points to form a trend line unique to this particular race.

Those call values, in seconds, are: 47.80, 98.40, 123.60 and 160.80 for 4, 8, 10 and 13 furlongs.

The final trend equation is $\hat{y} = 100.50\,x - 2.27$ for distances expressed in consecutive *fractional miles* equivalent to one through thirteen furlongs. Recall that $\hat{y}$ is the predicted y value (in seconds) from the regression equation for any given value of **x**, the distance.

For predictions in furlongs, the resulting equation is $\hat{y} = 12.56\,x - 2.27$. Readers should verify that these equations give exactly the same time values, $\hat{y}$, for their respective distances, provided that '**x**' is entered in proper units—decimal miles or furlongs, as appropriate.

The times, in seconds, for each consecutive furlong, regardless of the distance units, are: 10.29, 22.85, 35.42, 47.98,

Chapter Four

60.54, 73.10, 85.67, 98.23, 110.79, 123.35, 135.92, 148.48 and 161.04.

Since the COD for the prediction equations is a nearly unbelievable 0.999973 and the standard error of estimate is 0.300097, one expects such an *extremely good* fit of the resultant regression lines to the data.

Indeed, when the error is calculated between the values obtained from the regression equation and the actual official clocked times at the *four standard* call points, one obtains (in percents): 0.3719 at the quarter mile, -0.1747 at the mile, -0.1997 at the mile and one quarter, and 0.1498 at the finish.

These errors are so minute as to essentially constitute non-error.

Final Sophomore—and Career—Anomaly

Anyone remotely familiar with Man o' War's 21 career races knows that the only one in which he probably extended maximum effort for maximum distance was the Dwyer Stakes.

It was the fifth race of his three-year-old season and the fourth at Aqueduct on July 10, 1920. The distance was nine furlongs, and the track was rated fast.

The "field" consisted of John P. Grier, the only horse to oppose Man o' War. John P. carried 108 lbs. to Man o' War's 126.

At least this race had the saving grace of not being merely staged to prevent a walkover. In fact, it is probably fair to state that John P. Grier startled most observers by sustaining a real challenge to Man o' War from the start until near the finish.

Although Man o' War is listed as leading at each call point and winning by one and one-half lengths, the outcome was not decided until the last strides.

Man o' War's final time was 1:49.20. Since John P. Grier trailed by just 1.5 lengths, it is easily calculated that he finished in 1:49.42 s, just 0.22 s after the winner.

Man o' War's comment line states: "Hard ridden, drew away." (Daily Racing Form 28)

Calculating the time per furlong for Man o' War's 11 sophomore races shows that the Dwyer Stakes actually ranks as his second fastest effort. He averaged 12.13 s per furlong for the Dwyer and 11.975 s per furlong for the eight-furlong Withers Stakes, the second race of his sophomore year.

It is logical to think that the Withers time per furlong is consistent with Man o' War's general trend as a three-year-old since he carried just 118 lbs., as opposed to carrying 126 lbs. in the Dwyer Stakes over an extra furlong.

The combination of speed ratings and track variants for these two races show the typical sporadic and somewhat contradictory information that prevails in Thoroughbred racing and which makes true comparisons particularly difficult.

For the Withers, the SR-TV combination for Man o' War was 104-10. For the Dwyer it was 101-08. These numbers mean that Man o' War broke the existing track record for the Withers by 4/5 of a second, or 0.80 s on a day when the Track Variant, the *average of the speed ratings* for similar races at the track on that day, was 10—a full two seconds slower than the standing track record.

For the Dwyer Stakes, Man o' War broke the track record, going full out, by 1/5 second, or 0.20 s on a day when the Track Variant was 08—implying that similar races that day averaged 8/5 or 1.60 s slower than the track record.

Chapter Four

The conclusion must be, though the results be cursorily reviewed, that Man o' War performed better on a track that was presumably slower when he ran the Withers Stakes.

Considerations like this give some people an uneasy feeling about using speed figures, since they are based upon considerations of both Speed Ratings and Track Variants.

For now, however, it will do to summarize Man o' War's life after October 12, 1920, the day of his final career start. On that day at Kenilworth Racetrack in Windsor, Ontario, he defeated Sir Barton in a famous match race of 10 furlongs.

Much is made of the four-year-old Sir Barton, a Triple Crown Winner of the previous year, being beaten by seven lengths.

However, consistent with the many vagaries of Thoroughbred racing and by factors too subtle for many mortals to appreciate, Sir Barton was not in perfect health, due mainly to hoof problems. Additionally, since he was a year older than Man o' War, he had the dubious privilege of carrying 126 lbs. to Man o' War's 120.

Man o' War's combination of Speed Rating-Track Variant was 132-03 for the match race. His winning time of 2:03.00 thus beat the standing Kenilworth Track record that day by 32/5 seconds, or 6.40 s. This means that the track record, before that race, was 2:09.40 s.

Since Sir Barton crossed the finish line seven lengths back, this means that his time was 1.04 s slower than Man o' War's, or 2:04.04. This means, in turn, that Sir Barton, in losing, *also beat* the standing track record by 5.36 s—on unsound hooves and toting 6 pounds more than Man o' War.

It is interesting how few people mention this latter, but unquestionable, fact.

Life after Track Glory

As noted previously, Man o' War raced 11 times in his second, and final, racing year. Of those 11 races, he set records in eight. His earnings also surpassed Domino's, topping out at over $249,000. (Hollingsworth 128)

Having little or nothing more to prove, Samuel D. Riddle, Man o' War's owner, decided to withdraw him from further racing when he determined it likely that Man o' War would face high impost assignments if he raced as a four-year-old.

On January 28, 1921, Man o' War was galloped once around the Kentucky Association track in Lexington—the only time that he "ran" in Kentucky. From there he was sent for stud duty to the Hinata Stock Farm operated by Elizabeth Daingerfield.

In two breeding seasons at Hinata Man o' War sired 64 stakes winners, one being War Admiral who was an undefeated champion and Triple Crown winner at age three.

These duties completed, Man o' War was moved one final time in his life—to a place whose name, Faraway Farm, somehow seems to convey the idyllic sense of where so great a champion should be allowed to pursue his final years.

It was at Faraway Farm that groom, Will Harbut, of nearby Maddoxtown, Kentucky entered Man o' War's life. (Anderson 4)

Mr. Harbut was an exceptional human being, just as Man o' War was an exceptional equine. The two quickly developed a very close bond, Mr. Harbut taking care, on a daily basis, of every need the great champion presented.

Chapter Four

Indeed, Will Harbut also became known for the artful stories he would tell visitors about how 'Big Red' handily won each of his races. And many famous visitors there were, such as actress-singer Jeanette MacDonald, for one.

When occasionally questioned about the Sanford loss, it is rumored that Will Harbut unhesitatingly said that must have been a fiction because he never saw it! Such was his quick wit and charm.

All lovely things eventually end because that is simply the way of life on earth. It took 15 years for the special relationship between the gleaming chestnut stallion and the humble groom to wane, but eventually it did.

Will Harbut suffered a stroke, from which he essentially never recovered, in 1946. On October 3, 1947 a fatal heart attack at last removed Mr. Will Harbut of Maddoxtown, Kentucky, from this earth. The patron saint of animals also died on that date in 1226. I, for one, enjoy this "coincidence."

Were it etched in granite, it would be by no means a meager epitaph— being known as caretaker extraordinaire and beloved friend of the greatest Thoroughbred runner ever foaled in Kentucky.

Less than one month after Mr. Harbut's death, on November 1, 1947, Man o' War seemed agitated while being led to his stall. He is said to have balked and stood a while looking down the driveway as though for something familiar. Once within— perhaps seeing in some sense, unobstructed now by the sun's glare—his old friend beckoning, he lay down to seek his arms. He never arose. (Anderson 4)

Each being, man and horse, lives forever in the emotions of Thoroughbred lovers, their special relationship particularly emphasized by Ivan Dmitri's poignant photograph which became the September 13, 1941 cover of Saturday Evening Post.

> They were a field of champions, and he was just toying with them.
> ~ Ron Turcotte

Chapter Five

Secretariat

The town of Doswell, Virginia, lies some 20 miles north and slightly west of Richmond, the Commonwealth's capital.

Within the town limits, about 2600 acres formed the boundaries of The Meadow, the original homestead of Charles Dabney Morris, expanded from the original home around 1808. (Tweedy 7)

For centuries before Mr. Morris and the other direct ancestors of the Chenery family saw or acquired the land—the general valley area lying south of Fredericksburg and north of Richmond, Virginia—its sole inhabitants were the Youngtamund and Pamunkey Native American tribes whose lives were sustained by the largess of the North Anna and Mattaponi Rivers. (Tweedy 7)

These streams formed, respectively, the south-western and north-eastern limits of the broad valley from which The Meadow would take its final form and where one of the greatest American Thoroughbred runners first saw light.

Mr. Morris was an ancestor of Christopher Chenery who repurchased the land in 1936, it having passed from family control during the days following the Civil War.

Mr. Christopher T. Chenery grew up poor, and he was driven to achieve in ways few men understand. He was, fortunately, able to succeed in his quest, and thus he would eventually reclaim the land which was essentially taken from his forebears as spoils of the Civil War. (Tweedy 15)

From the time of his reclamation of and re-investment in his birthright, 34 years passed before the event transpired that eventually led to the first Triple Crown of Thoroughbred racing in 25 years.

At 12:10 am on the clear and cold morning of Monday, March 30, 1970 in the Meadow's foaling shed 17A, a colt was born to the mare, Something Royal.

He was her third foal by the pre-eminent American sire, Bold Ruler. His coat was burnished chestnut red, and he had three white "stockings," a star and a stripe. He was very sturdy and large, and he elicited a happily surprised exclamation from his owner, Penny Chenery Tweedy upon first sight. (Wolfe 25)

That he was foaled at The Meadow and would remain within its lush borders until January 20, 1972 (Tweedy 39) cannot be said to have been due to the luck of the draw. Rather, it was due to the luck of the toss.

The toss in question was of a half-dollar piece. It was held during the autumn of 1969 in a private office at Belmont Park racetrack in Elmont, New York. It occurred because it was by then an established tradition between Christopher T. Chenery and Ogden Phipps, owner of perhaps the greatest sire in the history of American racing, Bold Ruler.

Chapter Five

Mr. Chenery was now quite ill, and so his daughter, Mrs. Helen "Penny" Tweedy substituted for him on this final time, at her decision, the toss would occur.

This coin throw would determine who would have their choice of foals from two of the Chenery mares—Hasty Matelda or Something Royal—that had been bred to Bold Ruler in 1968.

In essence, the right of winner's choice for first foal was the "fee" charged under a two-year breeding contract between Mr. Phipps and Mr. Chenery.

The loser had first choice of foal the next year.

As it happened, Something Royal had dropped a filly and Hasty Matelda a colt in the spring of 1969. Something Royal was now again in foal, but Hasty Matelda was not.

Consequently, the winner of this toss would get one foal and the loser would get two.

Mr. Phipps called "tails" and won, though hindsight lends distinct humor to that word. He thus got Something Royal's filly. Mrs. Tweedy got Hasty Matelda's colt and whatever Something Royal dropped the following spring. (Wolfe 10)

The world now knows that Something Royal's foal, arriving in The Meadow's foaling shed 17A that early March morning in 1970 was eventually named Secretariat.

Secretariat's lineage, as did Man o' War's, reached far back into the early pages of James Weatherby's General Stud Book of 1791. Secretariat traces his male line to and beyond the unbeaten Eclipse, foaled April 1, 1764, a great-great grandson of the

Darley Arabian. His tail-male line is, in fact, the second longest in the Stud Book, beginning with Place's White Turk, imported to England in 1657. (Mackay-Smith 158)

In the 206 years separating Secretariat and Eclipse, the genes controlling white markings added two stockings to Secretariat. Eclipse sported a single "stocking' on his right hind leg. He also bequeathed his chestnut coat plus a star and stripe to Secretariat.

The Races

Secretariat's maiden race of his juvenile season was at Aqueduct. He was entered at five and a half furlongs in the second race against eleven other horses on July 4, 1972.

Paul Feliciano, a rather inexperienced rider, was his jockey for that and the next race. Thereafter, Secretariat was ridden only by Ron Turcotte except, unfortunately although with good outcome, for the final race of his career, the Canadian International in Toronto, Ontario, Canada where Eddie Maple raced him to victory by six and a half lengths.

The maiden race was disturbing, but it could have had truly disastrous consequences. Leaving gate 2, Secretariat was badly jostled about three strides out.

At that point a horse named Quebec, the gate-4 starter, swerved left and crossed in front of Strike the Line, the gate-3 starter. In so doing, Quebec ran squarely into Secretariat's right shoulder, knocking him leftward into Big Burn who had started from gate 1.

Secretariat was nearly knocked down. Only his inherent strength and athleticism kept him upright. His stamina and heart

Chapter Five

kept him contending, to the great surprise and awe of seasoned handicappers such as Davidowitz. (3).

He finished fourth by one and a quarter lengths, coming from nearly nine lengths back at the half-mile call point in a mere five and one half furlong race. It was the only loss of his juvenile season, and fourth was the farthest back he would ever finish in his career.

Even with the unbelievably bad start and the inexperienced rider he failed to place by only a nose to Fleet 'n Royal. Herbull was the winner and Master Achiever took second. Few Thoroughbred fanciers now remember those horses.

The remaining eight races of Secretariat's juvenile season went relatively smoothly. The first two were at Aqueduct, followed by three at Saratoga; then came two at Belmont Park, one at Laurel and the final juvenile run at Garden State Park.

Secretariat won his last eight juvenile races by an average winning margin of slightly over 3.8 lengths.

Even a cursory glance at his DRF past performance lines discloses a general style of running that he and Ron Turcotte established early. That is, he generally lagged the field at the first call point by fairly hefty margins. By the second call point he cut his lag distance by more than half. By the third call point he was *always* leading and, naturally, he then maintained that lead and crossed the finish line the winner.

His average cumulative lag at the first call point by his final juvenile race was 6.44 lengths. His greatest lag at that point was 10 lengths in the Laurel Futurity, while his shortest was 3.75 lengths in a July 31 allowance race at Saratoga.

It is all the more remarkable that, despite these major lag distances over a significant fraction of each race, Secretariat's average winning Speed Rating was 94 and his typical comment line states that he won either "handily," "easily," or "ridden out." The first two descriptors certainly suggest that he was not running hard, and the final one does not indicate an obviously determined effort. (Daily Racing Form 279)

Secretariat's mode of running thus distinctly contrasts with Man o' War's 10 juvenile efforts. In those, Man o' War led from start to finish in five races, held second place in three at the first call by an average slightly less than one length, was third in one, the concluding Futurity of his first season, by 1.5 lengths, and was fourth in his single losing effort in the Sanford by 1.5 lengths.

Unlike Man o' War, Secretariat had no outstanding anomalies about his juvenile year. The unfortunate consequences of his maiden race probably qualify more as bad racing luck than anomaly – as might his disqualification in the Champagne Stakes.

It was in his second racing year, 1973, that anomalies seemed to occur consistently and for various nearly unfathomable reasons.

It seems fair to say that, as great an overall career as Secretariat crafted, with a trace more common sense or guidance he would have truly registered a career for the ages.

The facts of his sophomore year are now presented unadorned. Presumably, readers will interpret them more within the perspective of what they could have been rather than what they were—even as impressive as Secretariat's Triple Crown efforts remain.

Chapter Five

Secretariat's Sophomore Season

It began on March 17, 1973 at Aqueduct Racetrack in New York. It was thirteen days short of Secretariat's chronological third birthday or foaling date. However, Thoroughbred conventions as they are, he was considered officially three years old on January 1 of 1973, as were the other 24,000-plus colts and fillies of his foal crop.

On that mid-March day, the seventh race at famed Aqueduct racetrack was run on a track rated 'sloppy.' Although all such ratings are relative, this is the fourth worst of nine rating levels assigned to dirt tracks by Daily Racing Form (DRF) in past performance records, the poorest being 'cuppy.' (DRF 443)

The race was a seven-furlong sprint—the longest of the sprint distances—and five horses besides Secretariat were entered. Secretariat drew gate 4.

It was the Bay Shore, and it was at grade-level 3, or G3. As such, and in retrospect, fans had scant reason to think, based on his juvenile efforts, that Secretariat would be remotely challenged by any in the field.

They were not surprised. He won by 4.5 lengths in what was described by DRF as a 'mild drive.' His time, without really trying, was 1:23.20. That translates to a fairly nifty 11.89 s per furlong, or 55.53 ft/s. This time/speed combination gave Secretariat an 85 Speed Rating with a Track Variant of 17—distinctly slower than average, irrespective of actual field quality.

The variant's magnitude implies that the combination of track surface condition and the horses' general class level that day were below par 100 by 17/5 of a second, or 3.4 s. Par is symbolized by '00' in Track Variant format.

What this generic description does not reveal is that Secretariat was severely jostled by a horse named Torsion about a half furlong from the gate. He was then walled in near the rail due to Ron Turcotte's misjudgment, as later admitted, which could have cost the win.

With only a furlong remaining, Secretariat found an opening and then left the other horses behind as he charged to the wire, 4.5 lengths ahead of Champagne Charlie and another 2.5 over Impecunious.

All such considerations noted, Secretariat's closing move leant an auspicious start to his sophomore year.

Secretariat's second race of 1973 was again at Aqueduct in a one-mile run called the Gotham Stakes. Its level was one step up—G2.

If one reviews Secretariat's sophomore races, a logical progression is evident in the grade levels and distances. This speaks of sound training and planning. However, when the sophomore year anomalies are discussed, at least a faint shadow lingers over his management.

He does, however, logically progress from G3 to G2 and then to G1 through successive distances of seven, eight and nine furlongs. Specifically, the races are the Bay Shore, Gotham and Wood Memorial.

He then logically progresses through the Triple Crown races—the Kentucky Derby, the Preakness and the Belmont, all G1 and having distances of ten, nine and one-half and twelve furlongs, respectively.

Chapter Five

He will win eight of his eleven remaining sophomore races after the Bay Shore. His losses in the Wood Memorial, the Whitney and the Woodward Stakes, the three Ws, constitute the anomalies for this chapter's discussion.

In the Gotham Stakes, Secretariat ran one mile on a fast Aqueduct track in 1:33.40. His time was 1.20 s off the world record which was set, and still stands, by Dr. Fager in 1968 at Arlington Park. (ARM 911)

Secretariat's Gotham time tied the track record for the distance and thus earned a speed rating of 100. It was on an April 7 day when the track variant was 08. This means that the typical race of *comparable* distance that day at Aqueduct had an average speed rating 8/5 of one second, or 1.60 s, slower than the track record.

This, in turn, implies that the track was probably slower than average, although its nominal speed is listed as fast. Some undetermined part of the Track Variant is attributable to the class of the runners. It is a distinctly irritating number—hinting at but not providing information accurately defining actual track speed.

The Gotham Stakes, like the Bay Shore Stakes, had a six-horse field. Secretariat started from gate 3. Flush was in slip 4, and Champagne Charlie held position 5.

Secretariat bumped the gate's sides and wobbled a bit as he lurched forward, but then settled down nicely.

Dawn Flight ran the first quarter in 0:23.20 with Secretariat back about a length in 0:23.40, it being generally accepted in racing parlance, although not truly accurate, that one-fifth of a second equals one length.

From that point on, Secretariat was soon in command, although Champagne Charlie went valiantly after him, especially until the final furlong when Secretariat pulled away and won by 3 lengths in 1:33.40. Flush trailed another 10 lengths in third.

With two sophomore races and two wins under his girth, so to speak, Secretariat now rested, at least between workouts, for 14 days in preparation for the Wood Memorial, also to be run at Aqueduct by Jamaica Bay.

Secretariat, perhaps to the delight of many apprentice numerologists, ran both the Bay Shore and the Gotham as the seventh races of their respective days. He was again assigned a number-seven race for the Wood Memorial. However, any hopes such budding seers may have held to that point as a cachet of their profession were soon dashed.

Let the Anomalies Begin

William Nack, in what is probably the definitive biography of Secretariat, notes that on Thursday, April 19—just two days before the Wood Memorial—Secretariat's exercise rider, Jimmy Gaffney, noticed that the big red horse did not seem quite right for his mile workout. (274)

He told this to Eddie Sweat, Secretariat's groom.

The next day, Gaffney again tried to work Secretariat. Again, Secretariat did not seem to be himself, and he was blowing, a distress signal, when they returned to the barn.

Chapter Five

A workout under Turcotte on April 17 also found Secretariat not responding to requests for effort.

Due to a recent death in trainer Lucien Laurin's family, Gaffney did not tell him a potential problem had surfaced.

Not until the morning of the Wood Memorial did New York Racing Association examining veterinarian, Dr. Manuel Gilman, note an abscess on the inside upper lip when he was looking for Secretariat's identifying tattoo. (Nack 274)

Perhaps the most crucial element of this situation was that apparently no one among those who knew of the abscess told jockey Ron Turcotte.

Therefore, he expected Secretariat to respond as in the previous races.

The result was, as Turcotte later noted, that Secretariat kept throwing his head when asked to take the bit. (ESPN) My feeling is that, even though he was not explicitly told of the problem, Ron Turcotte was an experienced enough jockey to have immediately sensed trouble by Secretariat's head throwing. Ostensibly he was not concerned and tired to ride Secretariat normally.

Secretariat finished third in the Wood Memorial, four lengths behind winner Angle Light owned by Canadian Edwin Whittaker. It proved slightly embarrassing, to say the least, that Lucien Laurin was also Angle Light's trainer and that he had assured Mrs. Penny Tweedy repeatedly not to worry, that Secretariat would win.

As events worked out, Ron Turcotte and Lucien Laurin both experienced good doses of Mrs. Tweedy's wrath, and probably justly, over the loss.

Not until Derby week, while Secretariat's entourage was preparing him for the Louisville race, did Ron Turcotte hear, second hand, what was undoubtedly the problem in the Wood Memorial.

He was in the jockey's room at Aqueduct when a friend who was an official of the New York Racing Association told him about the abscess and that it was discovered on the morning of the Wood Memorial. (Nack 297)

Although Turcotte was highly relieved that the loss was not his fault, the entire incident comes close to being ludicrous, considering Turcotte's experience and that Lucien Laurin was normally a stickler for Secretariat's health, had he known of the abscess, and about giving Turcotte handling instructions.

In fact, in each of Secretariat's losses during his final year of racing, factors related to health intervened between him and victory as they never did for Man o' War.

Perhaps these may all be classified as "racing luck." However they be categorized, they—and not some innate shortcoming of Secretariat's, were the only factors separating him from a record at least as impressive as any in racing.

Between May 5 and June 30, 1973, the big red horse from The Meadow of Doswell, Virginia more than compensated for any ill fortune the Fates previously dealt.

In succession, he not only won the Triple Crown, becoming the first horse in 25 years—since the great Citation—to do so, but he also set nearly unbelievable records in all three of its races.

Chapter Five

His Kentucky Derby come-from-behind—proving that he was back to his preferred running style—was the first clocking ever under two minutes in the first of the Crown's jewels. His time was 1:59.40. It required 77 years from the first 10-furlong Derby of 1896.

Only Monarchos in 2001 has come close, in the intervening forty years, to Secretariat's Derby record with a winning time of 1:59.97. (ARM 1173) Monarchos, however, did not achieve fame in the remaining Triple Crown races. Point Given won both the Preakness and the Belmont that year, with Monarchos finishing them in sixth and third place, respectively.

Contrary to comments made periodically by supposed racing buffs, Monarchos *is not* only the second horse to break two minutes in the Derby. Sham, in fact, required 0.21 s *less than* Monarchos by finishing 2.5 lengths behind Secretariat in the Derby. Sham's time, therefore, was 1:59.76, as compared to Monarchos' 2001 time of 1:59.97.

Monarchos' dream of Triple Crown glory was, unfortunately, short-lived.

The Preakness Stakes, held two weeks later, featured trainer Bob Baffert's Point Given crossing the wire first over ten other horses in the time of 1:55.51. By June 19, 2012 when the Maryland Racing Commission finalized Secretariat's official Preakness time (Associated Press 1), Point Given's time would be 2.51 s slower than Secretariat's record performance—1:53.00.

By that later date, the Maryland Racing Commission had enough evidence to declare that Secretariat had run his Preakness Stakes in the record time of 1:53.00. The timing error, suspected by many for 39 years as being incorrect, was finally vindicated.

The Daily Racing Form never accepted the "official" Pimlico time to begin with. Two of its independent clockers had timed Secretariat in 1:53.4, and that was the time DRF continued to publish in both their Champions editions of 1999 and 2005. (DRF 279)

It will be good to have all footnotes, asterisks or any other assorted corrective indicators removed from racing records once and for all—or at least until another major timing system fails at some prestigious event.

Monarchos finished sixth in the Preakness, seven and one-half lengths back. His time was 1:56.63, calculated by a method to be described later.

Point Given again triumphed in the Belmont Stakes, winning in a time of 2:26.56. Monarchos finished 13 lengths back in third place with a time of 2:28.51, 4.51 s slower than Secretariat's record.

Secretariat's Preakness was a miniature of his Kentucky Derby—in the sense that although the distance was a half furlong less, the race developed similarly to the Derby and Secretariat defeated the valiant Sham by an identical margin of 2.5 lengths.

The field was six for the Preakness rather than 13 as in the Derby. Secretariat began in gate 3 at Pimlico as opposed to gate 10 at Churchill Downs.

Now that Secretariat's Preakness finishing time is settled and accepted by the Maryland Racing Commission as 1:53.00, it is easily calculated that Sham officially holds the second

Chapter Five

fastest Preakness time by virtue of finishing 2.5 lengths behind Secretariat. Sham's time calculates to: 1:53.36.

Three horses have, up to last year's decision, been credited as co-holders of the Preakness record. They were: Tank's Prospect since 1985, Louis Quatorze since 1996 and Curlin since 2007. The American Racing Manual of 2012 actually lists their times as slightly different, ranging from 1:53.40 for Tank's Prospect to 1:53.43 for Louis Quatorze to 1:53.46 for Curlin. (ARM 1249-1250)

This is possibly due to increased accuracy of the timing devices over the years. Otherwise, it is inconsistent, at best.

However, all such considerations are now academic because Sham's time is faster.

It is unfortunate for Sham that he was foaled in the same crop as Secretariat. Had he been from a crop one year different in either direction, or vice versa regarding Secretariat, he possibly would have been the second most heralded equine runner of the past century's latter half.

An 8.5-furlong stakes, begun in 2001, now bears his name. It is a G3 event, which is nearly an insult to his ability, in my opinion. He was, indeed, an exceptional runner. The spirits of all Thoroughbred competitors deserve much better, no matter the number of their earthly accolades—and regardless of to whom he was second.

Sham, in fact, probably should not have run in the Belmont against Secretariat. That he did rests largely upon the shoulders of his trainer, Frank "Pancho" Martin.

Man o' War & Secretariat

According to Nack (380), Mr. Martin could never reconcile himself to the fact that Secretariat was probably faster, over any distance, than was Sham, even after Sham was convincingly beaten in both the Kentucky Derby and the Preakness.

Sham presented, to those of any emotional sensitivity, a dismal spectacle as he literally disintegrated as a runner during the final half of the Belmont Stakes.

Nack gives Secretariat's twelve splits, one per furlong, during the Belmont Stakes as follows: 12.20, 11.40, 11.40, 11.20, 12.00, 11.60, 12.20, 12.20, 12.00, 12.80, 12.20 and 12.80. (403)

The sum of these values is 144.00 s and their average is 12.00 s, as they must be. What is significant about this performance, in addition to it being a still-standing world record for the distance on dirt, is that Secretariat broke both his now-accepted Preakness record and his still-standing Kentucky Derby record at the same respective distances within the Belmont alone.

His times, applying the Belmont splits to the 9.5-furlong Preakness and to the 10-furlong Kentucky Derby were: 112.60 s and 119.00 s, respectively.

Those times are each 0.40 s less than he ran the individual races. Even more impressive about all this is Ron Turcotte's recorded comment that Secretariat was running easy during the Belmont. (ESPN)

Secretariat may have broken the 'then' unofficial world records at nearly every furlong point or pole (Tweedy 128), but he did not beat Dr. Fager's still standing 1968 record time for 8 furlongs, and only his times for 11 and 12 furlongs, his

Chapter Five

own Belmont record, surpass the 2011 world records for those distances.

His accomplishment, however, still stands as the most singularly impressive Thoroughbred performance in the sport's history.

After the spectacular triumphs of the Triple Crown, a kind of lethargy hit Secretariat's entire camp.

He went "four-for-six" in his final half-dozen career track encounters. It would have been a stellar performance for Ted Williams in a baseball double-header, but not for Secretariat on the track. Relativity works strangely in that regard.

Big Red's Final Anomalies

The six races capping Secretariat's career were, in chronological order: the Arlington Invitational at Arlington Park near Chicago, the Whitney Handicap at Saratoga, the Marlboro Cup Invitational Handicap at Belmont Park, the Woodward Stakes at Belmont Park, the Man o' War Stakes at Belmont Park and the Canadian International Stakes at Woodbine, Toronto, Ontario.

The last five of these six races were for three-year-olds and up. The final two races were on turf, and these were the sole turf performances of Secretariat's career.

The Arlington Invitational

The Arlington Invitational was a 9-furlong race. It was run

June 30, 1973, 21 days after the Belmont. The track was rated fast. The field was small, having only Secretariat and three other colts.

Secretariat was well rested for this event and the results more than displayed his condition. He stayed about twenty feet wide of the rail on both turns (Nack 409) and still won by nine lengths over My Gallant. It was then a neck to Our Native and 17 lengths more to the field. (DRF 279)

By finishing the nine furlongs in 1:47.00, Secretariat missed breaking Damascus' track and American Derby record (1:46.80) of August 5, 1967 by one-fifth of a second.

However, as many forget or overlook, Secretariat actually ran about 124 feet further in 1:47.00 than the nominal race distance by staying twenty feet wide of the rail on both turns. He actually ran at least 6,064 feet versus the nominal race distance of 5,940 feet, since the nominal distance around each turn at Arlington Park is one-quarter of a mile, as scaled from a satellite view. (Google 1)

It is easily calculated that his average speed was 56.67 ft/s for the longer distance. Dividing the nominal race distance by his "trip" speed shows that Secretariat ran the equivalent of nine furlongs in 1:44.81. That time beats Damascus' record by essentially two seconds.

When one considers that Secretariat's comment line for the Arlington Invitational states "easily," while Damascus' corresponding line states "ridden out," (DRF 202) this performance is impressive and probably ranks with the Triple Crown efforts.

Chapter Five

The Whitney Stakes

Secretariat was then transported to New York's Saratoga racetrack for the Whitney Stakes, scheduled for August 4.

Ron Turcotte noted more than once that Secretariat was not working out normally during July in preparation for the Whitney. Nonetheless, the decision was to run him—another anomalous decision. Does the term déjà vu seem appropriate?

During the race, he appeared dull and not fully responsive. He hooked Onion once in the stretch, but Onion outlasted him by a length with Rule by Reason another half length back in third.

The next day Secretariat was running a temperature. His veterinarian said that he probably had a virus before the race.

This loss was avoidable simply by not running him, but considerations of network television pressure prevailed. My grandmother Reid would have dusted off one of her favorite sayings about this situation: "The almighty dollar prevailed!"

The Marlboro Invitational

It was not until later in August that Secretariat began showing signs of his true form. He posted some exceptional workouts just in time for the running of the Marlboro on September 15 at Belmont Park. (Nack 411)

Secretariat's stable-mate, Riva Ridge, was also entered in the international field of seven. Another, to-be Hall of Fame jockey, Eddie Maple, rode Riva Ridge.

As would be expected of a healthy Secretariat, the Marlboro proved to be little contest. Secretariat won by 3.5 lengths over Riva Ridge who was another 2 lengths ahead of third-place finisher, Cougar II. The comment line states "ridden out." (DRF 279)

Despite the connotations of the comment line—that his jockey was using his hands to urge him on—Ron Turcotte said after the race that he knew Secretariat was simply "toying" with the other horses and yet they were all champions in their own rights. (Nack 412)

The Woodward

Things did not develop so smoothly in the Woodward, held at Belmont Park on September 29. In fact, Secretariat finished 4.5 lengths back in second place, blowing heavily as he returned to the unsaddling area. (Nack 413) This, however, would be the final time he would not win.

Secretariat set the pace most of the way, but a horse named Prove Out finally overtook him near the eighth pole and went on to win. Prove Out's time for the 12 f on a sloppy track was 2:25.80. It was, therefore, nearly two seconds slower than Secretariat's record for that distance and served as obvious indication that Secretariat was nowhere near his top form.

Man o' War Stakes

By the time October 8 came around, Secretariat had again been training very well. Lucien Laurin was able to give him what he called the "zinger" workout as a final preparation for the race.

It was an extremely fast workout on the turf course: 1:09.00 for six furlongs. It was actually 0.80 s faster than his six furlongs in the Belmont, which still stands as the record for that distance in that race.

Chapter Five

Secretariat's chief rival in the seven-horse field was undoubtedly Tentam, a four-year-old standout American grass horse.

The two ran the first mile and a quarter in 2:00.00 flat and then Secretariat simply ran away with the race, winning by five lengths in a track record time of 2:24.80 and earning a speed rating of 103. (DRF 279)

The stage was now duly set for the finale to his illustrious, if somewhat checkered, career.

The Canadian International

On a chilly day of glowering, leaden skies in late and oft' melancholy October, the town of Toronto in Ontario, Canada awaited the spectacle of the American superhorse running on Canadian turf for his final race.

Secretariat shipped well for the trip, and his racing team held high expectations for him. It was ironic that Ron Turcotte had been set down just days previous for rough riding. Although he would help announce the race, he would not ride the horse who, by his own admission, did it all on his own.

Turcotte did, however, work Secretariat five-eights of a mile three days before the race in a time of 0:57.60. It was a full second faster than the track record and it astonished the Canadian clockers.

For the race itself, Eddie Maple—an excellent jockey and future Hall of Fame inductee—would guide Big Red around the thirteen-furlong circuit. This final run would also be the longest of Secretariat's career and on turf.

Eddie Maple had just ridden Riva Ridge for Mrs. Tweedy the previous day, October 27, in what proved a decisive defeat in the Jockey Club Gold Cup. He was nervous of possibly losing Secretariat's final race, but nonetheless was assured that he would be up.

Eleven other horses were in the running, the most talented probably being Kennedy Road.

Secretariat did not disappoint the majority of fans who essentially came just to see him run and to bid farewell.

Basically being kept on "cruise control" by Maple, Secretariat entered the far turn head-to-head with Kennedy Road. Then Maple asked him to move, and he suddenly shifted into that magical spare gear only he, and perhaps Colin, possessed—opening five lengths while negotiating the turn and then adding seven more for good measure before clearing the turn and heading home. He was then leading by twelve lengths.

This consistent ability to "shift into a higher gear" on demand was what recently prompted Ron Turcotte to realize that he might be riding the greatest horse of all time—an image similar to what induced Chick Lang, Pimlico's General Manager, to quip that Secretariat looked like a Rolls Royce in a field of Volkswagens during the Preakness.

Undoubtedly the big red stallion presented a surreal image that dank October day—the hot breath from his nostrils spurting in concentrated, tortuous vapor plumes into the chill, damp air as he thrust his nearly 1200-pound frame, rumbling at its unchallengeable pace, down the head of the stretch toward the wire.

He coasted to the finish, having sufficiently demonstrated to the field his innate superiority, and still won by nearly seven lengths. The crowd, quite naturally, erupted in approval and adoration.

Chapter Five

With that win and the unforgettable image given to the frenzied gathering, Secretariat said goodbye forever to the racing world. He would soon ship to Claiborne Farm at Paris, Kentucky for stud duty in fulfillment of his $6.08 million syndication agreement.

Old Champions Never Die

In the remaining 16 years of his life, Secretariat sired some notable runners and important genetic exemplars. These included: Lady's Secret, Risen Star, General Assembly, and a daughter, Terlingua, who eventually foaled Storm Cat.

Lady's Secret and Risen Star were good enough to be honored by inclusion in DRF's 2005 Champions edition for their career achievements. (DRF 331, 340)

General Assembly still holds the record in the prestigious Travers Stakes at ten furlongs, set in 1979 when he ran it in 2:00.00. (Stretchrun 2). Risen Star won the 1988 Preakness and the Belmont. He might well have taken the Triple Crown, but was walled in during the Kentucky Derby and finished third.

His times for the Preakness, 1:56.20, and for the Belmont, 2:26.40, are distinctly slower than Secretariat's. (DRF 340) Much of this illustrates just how fast Secretariat was, as opposed to highlighting any deficit in his son, Risen Star.

Among knowledgeable horsemen, however, Secretariat is valued more as the sire of champion-producing mares than for any other ability.

His daughter, Terlingua, sired Storm Cat, and Storm Cat is considered one of the greatest of modern sires. It was Terlingua that reputedly brought so much favorable attention to her trainer, D. Wayne Lukas, that he opted to train full-time. (Wikipedia 1)

Secretariat did not quite produce a Triple Crown winner, as did Man o' War in begetting War Admiral. He came close in his son, Risen Star, but the "demon" racing luck intervened.

Neither great champion reproduced themselves, nor did they by any means produce a champion greater than themselves.

Producing a champion greater than oneself was held, by noted breeder Joseph Widener, to be the hallmark of the truly great champion. His argument was that Fair Play did just that in siring Man o' War, but that Man o' War did not equal the feat, even though he sired War Admiral. (McCarthy 104) I consider this dubious reasoning because genetic inheritance is much more luck than merit.

Likewise, Secretariat certainly did not reach that goal, although he proved that his sire, Bold Ruler, was truly great by the same reasoning applied in reverse.

Logical considerations argue that, genetically, there is a ceiling to ability among any species. If evolutionary theory is remotely correct, then species arise and mutate for survival in their given environment. (Meyer 8) This constitutes much of evolution's "survival-of-the-fittest" interpretation. Species do not need to "over-survive."

It is reasonable that, if the best of each species always produced a better descendant, then one could eventually reach an infinitely great or accomplished offspring. Not only would elementary mathematical concepts be thereby defied, but evolution's basic assumption would also be negated.

Suffice it to say that, human nature being as it is, the thinking species is never quite satisfied with anything and nearly always assumes it can improve something.

Chapter Five

Thus, a Gershwin song title becomes apropos: "It Ain't Necessarily So!"

Secretariat contracted laminitis, a serious hoof inflammation, apparently sometime during September of 1989.

It took scant time for the disease to place the great red horse's life in jeopardy. He was then just 19 years old. By October 3 Mr. Seth Hancock, owner of Claiborne Farm where Secretariat stood at stud since November 1973, realized that a decision to end that life was near. (Nack 447)

At the risk of offending the spirit of Abraham Lincoln—who, being the frontiersman and lover of creatures he was, probably would not have minded anyway—at 11:45 A.M. on the following morning, Wednesday, October 4, 1989, the incomparable Secretariat of The Meadow and of Claiborne Farm also became one for the ages.

He lies beneath an unpretentious stone marker—perfectly matched to his gentle personality—in the Claiborne Farm Equine Cemetery.

Many are the names of great Thoroughbreds carved upon the impassive, cold faces of other stones above other plots dotting the surrounding grounds. None, however, quite stokes the hearts and the imaginations and the emotions of visitors as does his.

It is truly right and just that none other probably ever will.

Prediction is very difficult, particularly about the future.
~ Niels Bohr

Chapter Six

Juvenile Year Correlations

Chapters 6 thru 10 present the essential data comparisons forming the crux of this book.

The comparisons proceed systematically throughout those chapters. That is, the comparisons that are simpler, more straightforward and less prone to bias, come first.

Using *correlations* between factors shared by both colts is a prime example. Correlations are easily understood and applied. They form a basis for further data exploration, and they often suggest additional useful comparisons.

After correlations, the *t-test* is applied to groups of factors which appear most highly correlated. The t-test indicates whether significant differences exist between the means of two samples.

After the t-tests, *simple linear regression* (SLR) is applied. This technique uses one independent and one dependent variable to produce a first-degree mathematical equation enabling one to predict running times for stated distances.

For all linear regressions herein, the independent variable is the *distance* run by a given colt, and the dependent variable is the colt's corresponding *time* for that distance. In a plot or graph of the resulting relationship between distance and time, time values are placed along the vertical axis, the *ordinate,* and distance values are placed along the horizontal axis, the *abscissa*, consistent with standard mathematical practice.

Last, *normalcy checks* are run using the Shapiro-Wilk algorithm on selected variables, and *simulated* "races" for the variables passing this test are run for pertinent distances. The simulations are based on Excel's Random Number Generator (RNG) function.

The above sequence of methods and comparisons helps develop interpretive skills for those new to even basic statistical applications. It also helps beginners feel comfortable and more confident in understanding unbiased statistical data comparison.

Basic Juvenile Year Data

During their two-year-old, or juvenile, racing seasons Man o' War ran ten times and Secretariat ran nine.

Man o' War's races were from June 6, 1919 to September 13, 1919. Secretariat's were from July 4, 1972 to November 18, 1972.

Data List 3 presents side-by-side comparisons of the race results. Distances are in both miles and furlongs; times are in seconds.

The results are given in chronological order from top to bottom. That is, the initial race of each colt's juvenile year is at the top and the final race is at the bottom.

This order reverses that of the DRF Champions 1999 and 2005 editions. (28) However, it seems more useful for this presentation.

Chapter Six

Data List 3

Man o' War		Secretariat	
Distance	Time	Distance	Time
0.625, 5f	59.00	0.6875, 5.5f	65.18
0.6875, 5.5f	65.60	0.75, 6f	70.60
0.6875, 5.5f	66.60	0.75, 6f	70.80
0.625, 5f	61.60	0.75, 6f	70.00
0.75, 6f	73.00	0.8125, 6.5f	76.20
0.75, 6f	72.40	0.8125, 6.5f	76.40
0.75, 6f	71.27	1.00, 8f	95.00
0.75, 6f	72.00	1.0625, 8.5f	102.80
0.75, 6f	73.00	1.0625, 8.5f	104.40
0.75, 6f	71.60		

The first, and most essential, point to note is that the only distance the colts ran, both *in common and for multiple times*, during their juvenile year, was six furlongs or 0.75 miles. The data for these distances are highlighted.

This means that the most accurate direct comparison at age two for both colts is exclusively for 0.75 miles. Other comparisons could be made, but they would involve some increased level of approximation, and that always introduces a greater element of inaccuracy or bias.

Therefore, unless there is absolutely no choice in the matter, one should directly compare only the results of running *equal distances* at multiple occurrences and *for equal ages and for the same sex* when judging the records of two or more horses.

Before addressing the comparisons, the correlations between ten selected race parameters which potentially affect the six-furlong times for each colt, are given.

Correlations with Time for Six Furlongs: Man o' War

Data List 4 includes ten parameters taken directly from the DRF past performance sheet for Man o' War's six-furlong races. (28) Explanations follow the numerical listing. The correlations are in high-to-low order, ignoring plus or minus signs.

Data List 4

Parameters	Correlation	Coefficient of Determination
Finish Margin:	0.5513	0.3039
Post Position:	- 0.5372	0.2886
Track Condition:	0.5372	0.2886
Field Size:	- 0.4309	0.1857
Impost:	0.4168	0.1737
Speed Rating:	- 0.3483	0.1213
Call Point 1:	0.2488	0.0619
Overall Gain:	0.1017	0.0103
Rest Period:	- 0.0296	0.0009
Track Variant:	- 0.0041	1.68E-05

Recall from the explanation of correlation in Chapter 2 that correlations are always real numbers between -1.00 and +1.00. If a correlation is positive it implies that as one of two variables (items being compared) becomes larger, the second variable also becomes larger, or vice versa. A negative correlation implies an *inverse* relation between two variables. Thus, if one variable gets larger, the second gets smaller, or vice versa.

Finish Margin refers to the number of lengths by which a horse either won or lost the race. The listed value, 0.5513, implies that higher finish margins are associated with higher finish times. This is contrary to expectation.

Chapter Six

Post Position is the number of the gate outward from the rail in which the horse starts the race. The gate nearest the rail is numbered one. The listed correlation, -0.5372, means that *higher* post numbers are associated with *lower* finishing times. This is contrary to expectation and may be due to the relatively small sample size of six in this case.

Track Condition refers to how fast or slow, *relative speed only*, the track is considered to be for each race on a given day. Daily Racing Form lists nine levels of track condition for dirt tracks. (417, 443) Man o' War ran all his career races on dirt. For Data List 4 correlations, the relative speeds such as *fast* through *cuppy* were assigned, or *coded*, numerical values ranging from 1 through 9, respectively. The listed correlation, 0.5372, is between the sequence of coded track-condition values and the corresponding race times. The positive value is consistent in that as track condition deteriorates, and thus is assigned *higher* coded values, the finish time also *increases*.

Field Size indicates how many horses were entered in the race. The correlation listed, -0.4309, is contrary to expectation. It implies that larger fields are associated with faster, or lower, finish times. More interference is expected for larger fields.

Impost means how much weight the horse was assigned to carry. It includes the weight of the jockey plus any tack, including the saddle, blankets, or other items. The correlation value, 0.4168, is consistent because it implies that the more weight carried the longer is the expected finish time. This, however, is an equivocal parameter.

Speed Rating refers to how many *fifths of one second* the horse finished the race, either faster or slower, than the track record held for the *past three years* for the given distance. Thus, if a horse *tied* the current track record, his Speed Rating is given the numerical value 100, also termed 'par.' For each fifth of one second that he either falls short of or beats the record, he is penalized one Speed Rating point or awarded one such point, respectively. (Ainslie 216)

For example, if a horse finished three-fifths of a second *slower* than the existing track record, his Speed Rating is 97. That is, it is 100 − 3. If he beats the track record by two full seconds, or *ten fifths* of one second, his Speed Rating is 110. That is, it is 100 + 10.

The correlation given, -0.3483, agrees with expectation because it implies that the higher the speed ratings, the lower were the finish times.

Call Point 1 means the first point in the race for which a time and a *margin*, the number of lengths a horse is either leading or behind at that point, are given by Daily Racing Form. For all six-furlong races the first call point is set at one-quarter mile, or two furlongs. The listed correlation, 0.2488, is consistent in that it implies that the finish time becomes longer the longer the time it takes for the horse to reach the first call point.

Overall Gain refers to how many lengths the horse made up, if he was behind the leader, or if he increased an already established lead, between call point 1 and the finish. The listed value, 0.1017, is inconsistent in that a larger margin gain from Call Point 1 to the finish should be associated with faster (lower) finishing times.

Rest Period means the number of days between each race of a given racing season. These periods allow a horse rest, except for scheduled workouts. No number is assigned to the first race of a given year because an indeterminate number of days occur

Chapter Six

between one racing season and the next. This makes the concept slightly awkward to use.

The listed correlation, - 0.0296, indicates an inverse but *essentially random* correlation between Man o' War's number of rest days and the times in which he ran his following races. This meets general expectation but is somewhat equivocal, in that, up to a point, one expects more rest to better prepare a given horse for competition.

Track Variant is a numerical rating given to the *average* Speed Rating of all the horses running on the same track on the same day and, ideally, under the same track and distance (sprint or route) conditions. It is interpreted, as is the Speed Rating, in terms of fifths of one second relative to the current track record of the past three years.

Therefore, if a Track Variant is listed as 07, it means that the *average* Speed Rating that day, under the same track conditions, and ideally for either sprint or route distances, for all horses, was 93. This implies 100 – 07, or 93. As listed, the inverse correlation value - 0.0041 is not expected because it means that larger Track Variants are related to faster (lower) finish times. Unfortunately, Track Variant is an ambiguous measure. (Ainslie 216)

Correlation is important because its sign, either plus or minus, shows the direction of a given relationship. However, the square of the correlation, the COD (*Coefficient of Determination*), indicates what percent of change between the two variables is actually 'explained' by the associated linear regression. If the squared correlation is small, then a significant relation is not indicated.

Any *separately calculated* correlation between two data sets, such as exist for the correlations in Data List 4, is duplicated—if a linear regression is also run on the same data sets—as one of the essential ten output values shown in Appendix D.

For the ten correlations in Data List 4, certainly the top five show relatively moderate strength. Rest Period and Track Variant have nearly random values. The COD for Track Variant means 1.68 multiplied by the negative fifth power of ten. It is small!

Data List 5 is the equivalent list for Secretariat. As important as these two lists are for each individual colt, the related comparison of the *pattern of correlations* between Secretariat and Man o' War for six furlongs discloses facets of their individual running styles which are not apparent just from the raw data contained in their past performance charts. Data List 6 provides the comparison correlations.

Correlations with Time for Six Furlongs: Secretariat

Data List 5 presents ten parameters taken directly from the past performance sheet for Secretariat's six-furlong races. (DRF 279) Explanations are identical to those for Man o' War's parameters. The correlations are listed in high-to-low order, disregarding sign.

Chapter Six

Data List 5

Parameters	Correlation	Coefficient of Determination
Speed Rating:	-0.8386	0.7033
Track Variant:	-0.6934	0.4808
Impost:	-0.6141	0.3772
Post Position:	0.5766	0.3325
Field Size:	0.5766	0.3325
Call Point 1:	0.3363	0.1131
Rest Period:	-0.2774	0.0769
Finish Margin:	-0.0524	0.0027
Overall Gain:	0.0476	0.0023
Track Condition:	see text	--------

As was done for Man o' War's case, a brief explanation and interpretation for Secretariat's ten parameters follows.

Speed Rating and the remaining parameters have the same meaning, naturally, for both Secretariat and Man o' War. Secretariat's Speed Rating correlation, -0.8386, indicates that, as his Speed Rating increased, his finish time for six furlongs decreased. This represents an inverse relationship between the two variables, as expected, with no further explanation required.

Secretariat's **Track Variant** correlation, -0.6934, also shows an inverse relationship with his times for six furlongs. This implies that as his Track Variant increased, his time decreased. This is contradictory to what is expected, and may be an artifact of his smaller sample size for these races—three versus six for Man o' War.

Secretariat's **Impost** correlation, -0.6141, implies that as his impost increased his time decreased. This is an inverse correlation and contradicts the expected relationship between the two values. The logic is that increased impost increases finish time. This is the ideal but it is inconsistent in practice. (Davidowitz 122)

Secretariat's **Post Position** correlation, 0.5766, implies that as his post position increased so did his finish time. This is the expected direction for these variables because higher post positions are further from the rail and horses need to run closer to the rail to conserve both distance and stamina. Thus, in theory, a horse must compensate for high post positions.

The **Field Size** correlation for Secretariat, 0.5766, indicates that as the field increased, his time increased. This is expected for these two variables. More horses would be expected to spell more potential interference and thus slower times.

The **Call Point 1** correlation, 0.3363, indicates that as Secretariat's time to the first Call Point increased, his final time also increased. This represents the standard expected relationship for these variables, but it can depend on the nature of a given horse's running style. It may be unexpected in his case, since he tended to show surprising acceleration in coming from behind and outflanking the field in many of his races.

A **Rest Period** correlation of -0.2774, implies that as Secretariat's time between races increased, his finish time decreased. This is the expected result of increased rest, up to a point, for any Thoroughbred runner.

The **Finish Margin** correlation, -0.0524, is in the expected direction, inverse to finish time. That is, the greater the margin by which Secretariat won, the lower was his winning time. Nevertheless, this relationship is small and approaches being random.

Chapter Six

The correlation value, 0.0476, for **Overall Gain**, that is, the number of lengths Secretariat gained from Call Point 1 to the finish, is also close to a zero or random value. Its squared-value, the COD, is 0.0023 and means that overall gain in lengths during the races only predicts about 0.23 percent of the changes in Secretariat's winning times.

Secretariat's **Track Condition** correlation for six furlongs has all coded values equaling '1' (by assignment) because he predominantly ran on *fast* surfaces in these races. In fact, eight of nine of Secretariat's juvenile races were on tracks rated 'fast,' thus receiving a '1' coding. His single race on a sloppy track was coded '4.'

The result is that a correlation coefficient cannot properly be calculated for six furlongs because division by zero occurs within the calculation. Excel returns the notice, #DIV0! indicating that situation and supported by the Data List 5 comment.

Coding the Correlation Sequences

Before proceeding to direct data analysis, it is informative to code the previous correlation sequences from Data Lists 4 and 5 and *then* correlate the coded sets. This can help determine how similar racing parameters affected each colt.

To do this, each parameter is assigned a separate coding integer from 1 through 10 based on, arbitrarily, the order of the parameters in Man o' War's list. The order is different in Secretariat's parameter list because corresponding parameters occupy *different* positions. This difference itself results in a new and unique correlation.

The resultant correlation of this "cross-coding" is -0.2531. The COD is 0.0641. The COD indicates that the *underlying pattern* influencing Man o' War's six-furlong races is related to Secretariat's by about 6 percent. It is, thus, essentially different.

This strongly *suggests* that the race parameter patterns most important to each colt's winning times essentially differ. One expects this from their known running styles alone.

Now it becomes more obvious what most influenced their running styles, given this correlation pattern, and the fact that they do not share a single parameter, *correlated in the same direction*, in their top five places. The three shared parameters have inverse signs.

Data List 6 gives side-by-side and high-to-low, ignoring sign, parameter comparisons.

Data List 6

Man o' War Correlations: 6 f		Secretariat Correlations: 6 f	
Parameters	Correlations	Parameters	Correlations
Final Margin	0.5513	Speed Rating	-0.8386
Post Position	-0.5372	Track Variant	-0.6934
Track Condition	0.5372	Impost	-0.6141
Field Size	-0.4309	Post Position	0.5766
Impost	0.4168	Field Size	0.5766
Speed Rating	-0.3483	Call Point 1	0.3363
Call Point 1	0.2488	Rest Period	-0.2774
Overall Gain	0.1017	Final Margin	-0.0524
Rest Period	-0.0286	Overall Gain	0.0476
Track Variant	-0.0041	Track Condition	0.0000

Highlighting in Data List 6 shows both colts' common parameters in its lower half. No other parameters common to the same half of the list have same-direction (sign) correlations.

Chapter Six

Call Point 1 time has the highest correlation-sign combination for both colts. However, it 'explains' just 6 percent of changes in Man o' War's six-furlong times and 11 percent of Secretariat's corresponding times. Rest Period 'explains' about 8 percent of Secretariat's six-furlong times but a negligible amount of Man o' War's.

Secretariat did, in fact, appear to have marginally more average rest period between six-furlong races, 14.33 days as opposed to 13.67 days for Man o' War, and a t-test between their corresponding parameter basically confirmed this judgment, having $\alpha = 0.44$ and thus being insignificant.

Overall Gain 'explains' about 1 percent of Man o' War's six-furlong times and a negligible amount of Secretariat's. Important or not, critical or not, these are facts the statistics indicate. Thus they cannot be challenged other than for their actual meaning and impact on performance.

This extended correlation study contrasting Man o' War and Secretariat has implications from several perspectives.

The single most important parameter for Man o' War relating to his six-furlong finish times is Finish Margin. The correlation for that parameter, 0.5513, accounts for roughly 30 percent of the changes in his finishing times, when it is squared to form the coefficient of determination, or COD.

Track Variant is the least important parameter for Man o' War regarding his six-furlong times. Its value, - 0.0041, translates to an extremely small coefficient of determination given in scientific notation by 1.68E-05. This implies that it predicts only 168 ten millionths of changes in his six-furlong finishing times. This would support its staunch critics, such as Davidowitz, cited earlier.

For Secretariat, the corresponding most and least important parameters are Speed Rating and Overall Gain. The correlation values of those parameters, -0.8386 and 0.0476, predict about 70 and 0.23 percent, respectively, of the changes in his six-furlong times.

In Secretariat's case, the relatively high correlation for Speed Rating can nearly be placed in the proverbial "no brainer" category. One expects, although this is sometimes contradicted, that high speed ratings equate to faster running times. It is, however, somewhat comforting to note that statistics does not always throw curve balls at interpretive efforts.

Man o' War's case for this parameter, nonetheless, remains anomalous.

It is appropriate now to compare the complete set of t-test values between Man o' War and Secretariat for the same ten parameters. Such comparison highlights those items which could be affected by era differences—if such differences actually existed.

Those parameters not showing significant differences cannot be interpreted as having been affected by inherent time differences, either actual or mythological.

Data List 7 presents the same ten parameters used for the correlation study. Any values of t either equal to or less than 0.05 imply a *significant difference* between the corresponding parameters for each colt. The significant values are highlighted.

t-Test Results for the Ten Race Parameters

The ten parameters are arranged by their significance values, least ($\alpha \geq 0.05$) to most, in Data List 7 in contrast with their correlation orders in the previous lists.

Data List 7

Man o' War and Secetariat t Values (α): Juvenile Race Parameters

Parameters	t-Test value
Post Position	0.4689
Rest Period	0.4438
Field Size	0.4365
Track Variant	0.3404
Track Condition	0.2582
Call Point 1	0.2132
Speed Rating	0.1393
Finish Margin	0.0949
Impost	0.0151
Overall Gain	0.0097

From the above results, it is apparent that only two of the ten race parameters differed significantly between Man o' War and Secretariat: Overall Gain and Impost. This is useful information. Remember that each of these tested differences between the colts refers to how a given parameter individually affected each colt.

A generalized discussion of six of the more interesting of these parameters now follows.

Rest Period, alpha level 0.4438, holds a non-significant position among the ten parameters. It is interesting that the nature of the t-test is such that the difference between the average rest times for the colts between six-furlong races, only about two-thirds of one day does not show as significant in terms of their final times.

The rest period time difference does not *seem* significant, but it is one thing to subjectively claim it is not and another to demonstrate it statistically.

One must look at the actual averages each colt had for rest time between races and then judge how much those differences made in their performances, if any.

Track Condition, alpha level 0.2582, tests at a less significant (further from 0.05) level than does Speed Rating. It would be logical to think the two were very close. Track Variant is actually the average Speed Rating for a day's races, whether those races are all sprints, all routes or some mixture thereof.

This α-value for track conditions between the two colts is non-significant because it is above the generally accepted 0.05 level.

That track conditions do not test significantly different between the two juveniles does not mean they were identical. In fact, it is one of the major contentions of those who object to comparing horses from different eras, that the tracks were much slower during Man o' War's racing years. It is not inconsistent to think that the tracks could have been much slower and yet receive similar Track Condition designators compared with

Chapter Six

Secretariat's tracks since the rating scale is entirely relative. This could account for the lack of significance found here.

A supposed Thoroughbred expert once exchanged an email with me in which he declared that if one automatically granted Man o' War at least four or five seconds per race due to track effect, then he might be properly compared with Secretariat.

It will be shown later that the inherent speed of Man o' War's tracks was not reasonably four or five seconds slower than those on which Secretariat ran.

There is currently *no absolute measure*—hence, no scientific measure—of track speed available for comparing various track surfaces. This point was thoroughly discussed in BG (Justice 142), and a variation of that discussion will be presented herein, as appropriate.

Readers are reminded, in cases of doubt, that Excel's inability to calculate a correlation between track conditions and finishing times for Secretariat's six-furlong races was due to their identical coding values for condition, since all but one of the tracks he ran were rated fast. They were thus all coded '1.'

However, 'fast' for one track on one day does not generally equate to 'fast' on another track on another day—or even for the same track between morning and afternoon. (Ainslie 115)

This fact about surface relativity and the 'forced' identical coding it generates automatically establishes a meaningless division by zero. Thus, Excel gives the error message, #DIV0! A true, scientific track measure would probably remove the circumstance of coding anomaly.

Calculating α between Man o' War's six track conditions for which he ran six furlongs against the three track conditions under which Secretariat ran the same distance, is a totally different application from correlation. It *does not* require an impermissible mathematical operation, and, therefore, a useful answer is given.

Call Point 1, alpha level 0.2132, holds sixth position on Data List 7. Given the distinctly different racing styles of the colts—breaking early for the lead and holding it versus staying back until well into the race and then skirting the field—of Man o' War, and Secretariat respectively, it is somewhat surprising that the t-test was not significant.

Man o' War, in fact, had several of his starts nearly recalled because he broke so fast from the barrier and rushed to the front of the field. His comment line for the Hudson Stakes, in fact, states: 'Broke through barrier.' There is, however, no recall indication.

Secretariat quickly learned to be wary of the start. He tended to stay back in the field when leaving the gate, perhaps due to the undoubtedly traumatic double-bumping which nearly collapsed him in his maiden run at Aqueduct.

Man o' War actually suffered his only loss, in the Sanford Stakes, allegedly because he was not prepared at the barrier when it lifted, and he never could make up the distance gap which followed because he was boxed in for much of the race.

Speed Rating, alpha level 0.1393, is the seventh list parameter. This α value approaches the borderline value for significance, usually set at $\alpha = 0.05$ for most statistical tests.

Man o' War's juvenile Speed Rating average was 89.83, whereas Secretariat's was 92.67. Certainly Secretariat's lighter imposts might be expected to account for some of this difference.

Overall Gain, alpha level 0.0097, is one of two parameters showing a significant difference for the two colts. When it is

Chapter Six

realized that Secretariat averaged 8.25 lengths gain from Call Point 1 to the finish of his six-furlong races, compared with 1.81 lengths for Man o' War, it is not surprising that this difference tests at a significant level.

Impost, alpha level 0.0151, has the second highest significance level of the ten parameters. This is not surprising because Secretariat's average six-furlong impost was 117.33 lbs. to Man o' War's 129.50 lbs. These figures do not include plate weights—about 0.90 lbs for a complete set of Aluminum plates as worn by Secretariat and 2.62 lbs. for the steel plates worn by Man o' War.

The calculated difference plate weights made to their respective racing performances is, however, negligible. It is estimated from linear regression to be of the order of 0.20 s.

Six Furlongs versus all Races

It is appropriate to end this section by asking a question and then proposing its answer. That answer is pertinent to a concern that simple linear regression, used in the following section, poses.

That is, only the six-furlong race results have thus far been discussed. It has been already stated that this distance was the only distance Man o' War and Secretariat raced in common, and also for multiple races, as juveniles.

However, linear regression uses the results of *all races* in a given season to form a broad view of a horse's ability. What guarantee is there that the six-furlong races do not show distinctly different patterns than the remaining races a horse ran in a given year?

This is where the t-test again becomes important and useful. The question is answered by taking the *mean value* of each of the ten selected race parameters for each horse for *all remaining juvenile races* and then *comparing* those mean values *with the mean values of only the six-furlong races*. If the t-test discloses a significant result, it is possible that an intrinsic difference exists in the effect these parameters have for varying distances. If that is true, it argues strongly for placing less reliance on the results of linear regression.

Data List 8 presents the two sets of means for both Secretariat and Man o' War. Units are omitted for simplicity.

Data List 8

Parameter	6 f	Remaining	Parameter	6 f	Remaining
Call Point 1	23.18	23.20	Call Point 1	23.41	24.00
Field Size	8.00	5.50	Field Size	7.67	8.67
Overall Gain	1.81	3.06	Overall Gain	8.25	10.46
Impost	129.40	120.00	Impost	117.33	120.33
Final Margin	1.67	3.25	Final Margin	3.50	3.17
Post Position	4.83	4.00	Post Position	4.67	4.83
Rest Period	13.67	5.67	Rest Period	14.33	18.80
Speed Rating	89.83	87.25	Speed Rating	92.67	93.17
Track Variant	12.83	14.00	Track Variant	13.67	13.50
Track Condition	1.83	2.75	Track Condition	1.00	1.50

The result of the t comparison is both comforting and informative. It is comforting because it helps reassure us there is no significant difference between either set of mean values. The α value for Man o' War's set comparison is 0.46, rounded to two decimal places.

Chapter Six

The α value for Secretariat's set comparison is 0.48 when similarly rounded. The two values are essentially identical.

The near equality of the two α values, at minimum, alludes to the conclusion that the factors operating during the races of both colts, and which affected the finishing times, *were probably distributed evenly* across the parameters.

This conclusion is not "carved in granite," but is strongly suggested. It would be highly coincidental otherwise that two completely independent data sets from colts running fifty-three years apart, should compare so closely.

The other useful information comes from the F test. This is the test of variance between data sets which *must* be run before performing the t-test. The F test determines whether the *variances* of the data sets (the data spread or ranges) *differ significantly*. If they do, that determines whether a '2' or a '3' must be entered in the last field of Excel's t-test Function Arguments Dialog Box before the test is run.

The '2' denotes *homoscedasticity*, the academic-sounding term introduced earlier. All the term means is equal variance—which, in turn, means that the data spreads, or standard deviations, around the separate means are not significantly different.

A '3' in the final t-test field tells Excel to adjust its t calculations for *non-homoscedactic* data—meaning that the two colts had *significantly different* data spreads around their separate means.

It is suggested that readers use the preceding list and run both the F test and t-test for practice and to convince themselves they understand these procedures.

The next chapter explores the relationship between six-furlong times.

As a final check of similarity between the ten parameters of the six-furlong races, a t-test of Man o' War's versus Secretariat's parameters gave an alpha level of 0.50 to two decimal places. The two sets are not significantly different.

Before proceeding, it should be noted that, all time adjustments aside, 13 of Man o' War's 21 career races, essentially 62%, were run either at Saratoga or Belmont. Specifically, six were run at Saratoga and seven were run at Belmont.

This is significant because Ours devotes a full half page to discussing the improvements to both tracks, before Man o' War ran either, for the express purpose of making them fast.

Saratoga was modified in August 1918 and Belmont was similarly altered sometime before mid-May 1919. Man o' War ran his first juvenile race on June 6, 1919 at Belmont.

Ours notes that, after the alterations, records started falling so quickly at Belmont that it became known in some circles as the "rubber track." (58)

It is interesting that none of the Man o' War devotees I've read ever mention this fact.

Comparisons are odorous.
~ Shakespeare

Chapter Seven

Test for Significant Difference at Six Furlongs

It is appropriate now to compare the actual times of Man o' War and Secretariat for the only distance they ran in common as juveniles—six furlongs.

The *first such test* will be a t-test of their *raw times*—that is, the times as they stand in the record books without modification. This is the simplest t-test to run. We expect that a significant difference exists between these times. However, the difference is not due to any true era difference. It is due simply to the fact that Man o' War ran on somewhat slower surfaces then did Secretariat. He also wore steel racing plates, as opposed to the aluminum plates Secretariat wore. That, presumably, makes a slight difference.

Equitable adjustments to the times of both colts will be made and tested for effect. The rationale for these adjustments is now given.

Setting Boundaries for Man o' War's Time Adjustments

If reasonable boundaries are first determined within which Man o' War's adjusted times should logically fall, much confusion is later avoided. It is, fortunately, possible and relatively easy to set these limits.

We begin by noting that Man o' War's average and standard deviation for his six-furlong races are 72.21 s and 0.72 s respectively. Likewise, Secretariat's matching parameters for six furlongs are 70.47 s and 0.42 s, respectively.

A simpler version of the t-test formula, introduced in Chapter 2, can be used to determine these initial, critical boundaries. The boundaries are introduced first. The formula and its explanation are detailed in Appendix H.

Refer now to Figure 7, in which all numbered points represent seconds.

Figure 7

The logical boundaries for Man o' War's Time Adjustments: 6 f

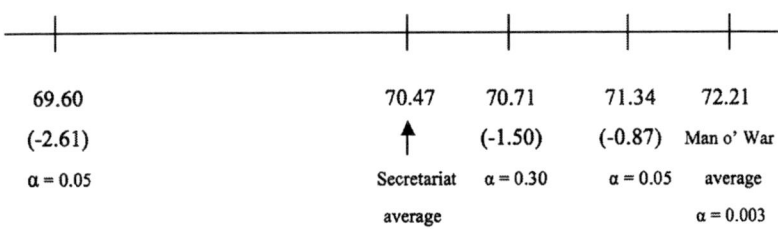

69.60		70.47	70.71		71.34	72.21
(-2.61)		↑	(-1.50)		(-0.87)	Man o' War
α = 0.05		Secretariat	α = 0.30		α = 0.05	average
		average				α = 0.003

Proceeding from *right to left* along the line in Figure 7, the following significance attaches to the listed numbers:

First Comparison: 72.21 s is Man o' War's *unadjusted* average time for six furlongs. A t-test of his original six-furlong times against Secretariat's (70.47 s) shows a significant difference at the 0.003 level. This is not unexpected, as stated previously.

Chapter Seven

Ten sets of 100 random simulations run on the raw data for the two *average* times resulted in Man o' War having 20 of 1000 faster predicted times than Secretariat. The actual sequence of faster times predictions was: 3, 2, 2, 2, 0, 1, 2, 3, 3 and 2 f.

Second Comparison: 71.34 s is the new average six-furlong time for Man o' War if 0.87 s is subtracted from *each* of his original times. The number 0.87 was calculated using the simplified t formula from Appendix H so that Man o' War's final six-furlong average time fell on the *upper border* between significance and non-significance when compared with Secretariat's average unadjusted time for six furlongs. Hence, the meaning of '$\alpha = 0.05$' beneath the parenthetical number.

The upper 0.05 boundary implies that 0.87 s is the *most* that can be subtracted from each of Man o' War's original times and still leave them significantly different (marginally) from Secretariat's.

Ten sets of 100 random simulations run for this average time adjustment results in Man o' War having 168 of 1000 times faster than Secretariat. The sequence of random results obtained was: 16, 21, 12, 18, 14, 20, 16, 11, 16 and 24.

Third Comparison: 70.71 s is the resultant average six-furlong time for Man o' War when 1.50 s is subtracted from each of his original six-furlong times. This number is considered the nearly proportional equivalent for six furlongs to Farley's (337) estimate of *almost* 2.00 s for eight furlongs. If Farley meant 1.9 s, the exact equivalent is 1.43.

Ten sets of 100 random simulations based on this average resulted in 397 of 1000 faster times assigned to Man o' War. As shown by the corresponding value, $\alpha = 0.30$, Man o' War's adjusted times are no longer significantly different from

Secretariat's unadjusted times. The two sets differ insignificantly at the 0.30 level, a fairly strong indicator of equivalency. The actual sequence of random results was: 32, 37, 34, 48, 38, 41, 55, 37, 30 and 45.

Fourth Comparison: The next number proceeding leftward, 70.47, is Secretariat's unadjusted average time for six furlongs. It also represents the average attained if 1.74 s is subtracted from *each* of Man o' War's original times.

This particular adjustment is interesting from the following perspective. That is, it suggests agreement with Farley's estimate, previously cited, that Man o' War's tracks were *almost* two seconds slower *to the mile* than were Secretariat's. Given the nature of randomness in simulating the results, the fact that 504 of 1000 such simulations show Man o' War with a faster time than Secretariat represents a nearly perfect 50-50 split.

Since the averages for both colts are identical for this level of adjustment, the only difference is their standard deviations. Therefore, it may reasonably be stated that differences in track effects were the primary cause of Man o' War's standard deviation of 0.72 s and of Secretariat's 0.42 s standard deviation.

The actual sequence of random values obtained for this adjustment was: 47, 57, 53, 51, 57, 51, 35, 45, 53 and 55.

Fifth Comparison: When each colt's z-score for six furlongs is referenced to the 2011 world record average for six furlongs, the result is equivalent to subtracting 1.71 s from each of Man o' War's times. This nearly equals the previous adjustment to Secretariat average time and is not surprising.

Based on how z-scores are calculated, they represent a kind of second average. Since the result is so near the previous value, it was omitted from Figure 7 to avoid cluttering.

Chapter Seven

However, it is interesting that the seemingly insignificant difference of 0.03 s between the two adjustments results in 506 of 1000 times for Man o' War being faster, vice 504 of 1000 for the previous adjustment. This difference is also insignificant.

The exact sequence of random values obtained for the -1.71 s adjustment was: 52, 55, 49, 41, 50, 46, 45, 58, 53 and 57. The z-scores are fully discussed in Chapter 8.

Sixth Comparison: The sixth, and final, test comparison was used to establish a lower boundary beyond which it is *unreasonable* to extend Man o' War's adjustments. This value was calculated from the t-test formula such that it gives Man o' War an edge *on the superiority side* of the probability scale.

That is, if 2.61 s is subtracted from *each* of his six-furlong times and those times are then checked against Secretariat's original six-furlong times via the t-test, then Man o' War is on the border of being *significantly ahead*, in terms of faster times, of Secretariat with an average time of 69.60 s to Secretariat's 70.47 s.

This adjustment assigns 871 of 1000 faster simulation times to Man o' War.

One implication of the 2.61s lower boundary is that it represents the maximum reasonable difference by which Man o' War's track speeds for six furlongs could have lagged Secretariat's—by definition.

That is because Man o' War's times for that level exceed the 0.05 significance level when so reduced, and *all further reductions* make them differ by even greater significance from Secretariat's. That would neither be fair nor a likely reality.

The sequence of random time values per 100 trials (favoring Man o' War) obtained for this adjustment level was: 87, 86, 83, 87, 87, 87, 85, 88, 89 and 92.

Man o' War & Secretariat

It is assumed that fair-minded Man o' War devotees will at least grudgingly agree that Secretariat and Man o' War were more closely matched than they were different regarding running ability. Therefore, one would expect neither colt to pass a hypothetical boundary when so doing would make him appear obviously faster.

Note also that, regardless of which adjustment level is used, the standard deviations of both colts remains unaltered. That is because the same amount may either be added or subtracted from each value within *any* set of data values and leave the set's standard deviation unaffected.

To demonstrate this, if necessary, consider an extravagant example that is easily obtained by subtracting 70 s *from each* of Man o' War's run times for six furlongs. The resultant values then range from 1.27 s to 3.00 s.

These outrageous times imply that his speed, at 1.27 s for six furlongs, was over 2,100 mi/hr—some 1300 mi/hr faster than sound ($\approx$ 741 mi/hr) at sea level and nine times faster than a typical Indy car!

Analogous to what the Germans call Superman—'Übermensch'—Man o' War then becomes 'Überpferd.'

However, one can still do this and see that the standard deviation is unaltered. It remains 0.72. And you thought statisticians lacked humor!

More important than humor now, if the standard deviations remain constant, one is assured that homogeneity of *variance* (standard deviation squared) is maintained. Therefore, the one-tailed t-test, for equal variance, can be run and a valid result obtained no matter what the adjustment. This version of the t-test is considered more reliable. (Schmuller 169)

Chapter Seven

Those who still wish to argue that an era difference existed between the two colts such that it rendered equitable adjustments and comparisons unattainable will not relish the conclusions of this section. Although unfortunate, it cannot be helped. It is the reality.

My second book on Thoroughbred data comparison, BG, acknowledges that a true era difference did exist, for instance, between aircraft of the first two World Wars.

The Fokker Dr.I of WWI and the U.S. Navy's F6F Hellcat of WWII were used as examples in BG. The latter planes were a product of drastically advanced technology, science, engineering and production skills developed during the approximately twenty years between wars. The *planes themselves* were dramatically changed across time and the F6F could not reasonably be compared with the Fokker regarding fighting abilities.

However, the *basic nature of the Thoroughbred did not change* in the fifty-three year span between Man o' War and Secretariat. The same gene pool was still in effect for the Thoroughbred equine. Time's passing may have bred *more* skilled runners. It did not basically alter the physiological or metabolic processes propelling those runners. (Cunningham 97)

Two main physical effects are basically all that separate Man o' War and Secretariat: possible *track differences* and, marginally, *impost* and racing plate differences.

Data List 9 recapitulates the ten random sequences of 100 simulations each upon which the results of the six previous comparisons were based. These results are given as reminders that random simulations fluctuate over fairly broad ranges. There is no single exact outcome, by definition, of such simulations. In this aspect, they are not flawed, contrary to some interpretations.

Rather, they reflect nature's inherent uncertainty, known to exist at the atomic level since 1901 by physicist Max Planck. (Anderson 47)

Data List 9
Random Simulations of the Six Time-Adjustment Methods

First:	3, 2, 2, 2, 0, 1, 2, 3, 3, 2	Total: 20
Second:	16, 21, 12, 18, 14, 20, 16, 11, 16, 24	Total: 168
Third:	32, 37, 34, 48, 38, 41, 55, 37, 30, 45	Total: 397
Fourth:	47, 57, 53, 51, 57, 51, 35, 45, 53, 55	Total: 504
Fifth:	52, 55, 49, 41, 50, 46, 45, 58, 53, 57	Total 506
Sixth:	87, 86, 83, 87, 87, 87, 85, 88, 89, 92	Total: 871

Unbiased Time Adjustments to the Raw Data

The previous section gave broad guidelines for appropriate adjustments. This section provides detail on *specific applications* of those guidelines.

The pertinent data for the six-furlong races of Man o' War and Secretariat are in Data List 10. Track abbreviations include: Aqu = Aqueduct; Sar = Saratoga; Bel = Belmont. Times are in seconds to finish. Speed Ratings, SR, and Track Variants, TV, are in Daily Racing Form format. Each number is referenced to fifths of one second either above or below par, as explained previously.

Recall that par equals 100, by definition, and is also the current track record for the past three years at the time the given race was run.

Chapter Seven

Data List 10

Man o' War				Secretariat			
Date	Track	Time	SR-TV	Date	Track	Time	SR-TV
7/5/19	Aqu	73.00	90-12	7/15/72	Aqu	70.60	90-14
8/2/19	Sar	72.40	90-13	7/31/72	Sar	70.80	92-13
8/13/19	Sar	71.27	95-09	8/16/72	Sar	70.00	96-14
8/23/19	Sar	72.00	92-08				
8/30/19	Sar	73.00	87-14				
9/13/19	Bel	71.60	85-21				

The first step taken, along the pathway of the forbidden comparison, is to perform a direct t-test of these *raw data* (unadjusted data) and determine whether or not a significant difference exists between them.

The six finish times for Man o' War are entered in an Excel array, and the three finish times for Secretariat are entered in an adjacent array. When the t-test previously described is performed, the value 0.003, truncated to three decimal places, is obtained.

This value is well below the standard 0.05 significance level. It indicates, with little surprise, that *a significant difference does exist* between the two sets of times.

Before performing this test, an F test is first run to determine whether or not the spreads or standard deviations of the two sets differ significantly. The F value 0.54 indicates that both colts had essentially equivalent standard deviations in running times for the distance since F exceeded 0.05.

Therefore, a *one-tailed test* was chosen, vice a two-tailed test, because simple visual comparison of the data showed that Man o' War's individual times for six furlongs were greater than Secretariat's. Recall that when the equal-variance condition holds between any two data sets, it is better to run a one-tailed test because that is generally the more accurate assessment of the difference between the means. See Schmuller (169).

The reader should verify these results for practice and further understanding, as in all critical comparison cases discussed herein.

Adjusting Running Times for Data Comparability

The logical question now arises as to *what equitable adjustment can be made* regarding possible track surface and impost differences, including plate weight, to data which made Man o' War's times compare unfavorably to Secretariat's.

In my previous book, BG, the need for a *standard* by which to judge the speed of race tracks was discussed. The analogy was made regarding *absolute zero*'s use as a temperature reference standard in the physical sciences and comparing other temperatures to that point.

Analogously, the *standard of track speed* to which both Man o' War's and Secretariat's times can be adjusted is the current North American Dirt Track Record for six furlongs, as published in the American Racing Manual (ARM) 2012 edition. (911)

The ARM lists the record for dirt tracks, as of 2011, as being held by Twin Sparks. It was set at Turf Paradise (AZ) on November 21, 2009. Twin Sparks was then six years old and carried 114 lbs. in establishing the record time of 1:06.49, or 66.49 s.

Chapter Seven

At this point in the analysis, age and impost are *not important factors* to consider. What is first needed is an absolute benchmark—one that will probably take a significant time to break, and, even if surpassed, will likely be surpassed only marginally.

In fact, since the same criterion will be applied exclusively to Man o' War's and Secretariat's specific races, it does not matter whether this benchmark is absolute in the sense that absolute zero, as far as is anticipated, is an absolutely unwavering constant within the universe.

This book makes one major supposition about Thoroughbred running ability—that it is genetically determined within the species, and *only the most highly gifted runners ever perform near the topmost level of ability*—whether by luck or by design on a given day.

Even if the horse setting the record never goes beyond that record to achieve popular acclaim or fame, he or she must have been running at least *reasonably near* the species minus-three-sigma limit to establish such a time. That, at least, is the philosophy.

By definition, if a horse has established a world record for any distance, especially the longer that record stands, it is arguable that the record represents near the apex of ability within the given species—analogous to absolute zero in physical science.

Stated differently, a world record may be provisionally viewed—even though it was set 92 years after Man o' War ran and 39 years after Secretariat ran—as the *lowest potential time for the distance* attainable by an equine.

These *a priori* assertions given, adjustments will be made to the running times of both Man o' War and Secretariat to "balance" them, first with respect to their specific track records and conditions when they ran, and, later, to the current world record of 2011 published in the ARM.

The foregoing perspective is equivalent to saying that the speed of light was always the fastest that anything in the universe could travel although millennia passed before humankind even existed to discover its actual value and use it. It says nothing more or less.

Proof is now offered to substantiate the claim of relative constancy of Thoroughbred ability across generations. See Cunningham's article for more information. (92)

The Uniform Distribution of Thoroughbred Ability: I

A general data sort was performed, and lists were derived, for the pertinent world record holders listed in the ARM for the common distances six through ten furlongs. Each age from two through seven was represented within the total group of 312 runners that attained their respective world records between the years 1991 through 2011.

Data List 11 shows the total number of runners, N, for given distances and the age ranges, AR, represented within a given distance record. Data List 12 gives the total number out of 312 representing a given age and the percent of all runners that age includes.

Chapter Seven

Data List 11: Distance

	6f	7f	8f	8.5f	9f	10f
N	48	54	54	50	52	54
AR	2-7	2-7	3-7	3-7	3-6	3-6

Data List 12: Age

Ages:	2	3	4	5	6	7
Total	4	70	126	72	29	11
%	1.28	22.44	40.38	23.08	9.29	3.53

These lists give evidence that, as in most areas of Thoroughbred data analysis, generalizations are hard-won. For instance, most horsemen agree that a colt or filly is not mature—at which time they earn the names 'horse' and 'mare,' respectively—until age 5.

However, ages two through four represent slightly over 64 percent of the top world-record holders for the given distances. Granted that the two-year-olds are represented only within the six- and seven-furlong sprint distances, this is still fairly remarkable.

The fact that four 'babies,' as two-year-olds are frequently called, can compete at this level with equines considered fully mature, indicates that Thoroughbred ability is dictated at least as much by species genetics as it is by training.

If a two-year-old were *truly unable* to compete at world-record level, regardless of the distance, then no amount of training would enable him or her to do so. In fact, attempting to train that which is yet unprepared for training would likely ruin the career prospects of an otherwise talented youngster.

To cite a pertinent example, Samuel Doyle Riddle—Man o' War's owner—was ostensibly so solicitous over Man o' War's welfare that he would not run him in the 1920 Kentucky Derby because he felt that the mandated 126-lb. impost was too much for a young three-year-old to carry for ten furlongs.

Thus, Mr. Riddle unwittingly established what may be the greatest irony in all of sports—that the greatest equine runner ever foaled in Kentucky never raced in that Commonwealth and, furthermore, did not have a chance to win the Triple Crown, even after the fact, as did Sir Barton and Gallant Fox.

The Uniform Distribution of Thoroughbred Ability: II

A second fact, related directly to the above age discussion, concerns the world-record linear regression for the same horses. Refer to Figure 8.

Figure 8

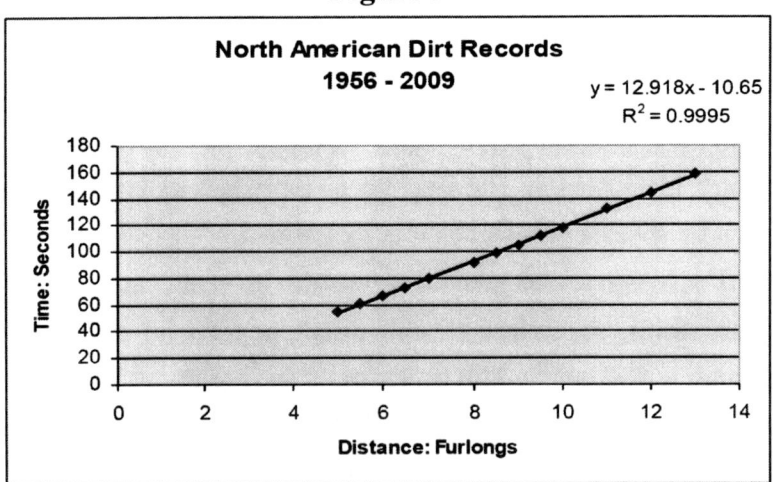

Figure 8 displays a *nearly perfect fit* to a set of 13 data points. The COD for the linear regression of world record times from five

furlongs through 13 f is 0.9995. This is the basic graph used as a reference standard for the "trapezoidal corrections" (explained later in this chapter) between Man o' War and Secretariat.

It is remarkable not only that this regression line is genuine rather than bogus, but also that it represents 15 (two ties) different horses running 12 different tracks under varied conditions.

No less remarkable is that their ages range from two through seven, that the records span 53 years—from the earliest held record in 1956 (Swaps, 13f) to the most recent 2009 records—and that such a uniform line is produced from such seemingly disparate circumstances.

One need not further gild the lily regarding this extraordinary line. It need only be said that, if Thoroughbred running ability is *not* uniformly distributed over at least three generations of horses, then in this instance someone forgot to tell the horses.

As was mentioned in GHA regarding individual linear regressions, it appears that Thoroughbreds, at least at the upper echelon of running ability, perform as though an internal autopilot is guiding there performances, so uniform do the graphs of their running times versus distance appear.

The point is now abundantly made. This and the preceding section give ample evidence that the postulate of *uniform Thoroughbred running ability across generations* finds support by real data. The world record regression virtually shouts it.

Therefore, it becomes increasingly unlikely that any era effect, in a true sense, separates Man o' War and Secretariat.

Man o' War and Secretariat ran on different track surfaces. So did the current world-record holders. Man o' War and Secretariat were separated by 53 years. The world record holders as a group are separated by 53 years. Man o' War and

Secretariat bore different imposts. So did the current world record holders—over a 19-pound range!

By now, one can almost picture a bemused Garfield rubbing his chin and chiding—"Go figure!"

As an aside, but also reminding readers of a previous comment, the world-record linear trend even justifies the concept that Thoroughbred running ability has been essentially unchanged since at least the time of Devonshire 'Flying' Childers in 1721.

His recorded time (6:40.00) for the measured distance of 3.80 miles, or 30.4 furlongs, fits nicely within the *current world record* distances and corresponding times. In fact, it is faster by nearly 10 s than the current world-record time for the lesser distance of 3.625 miles or 29 furlongs. That time is given in the 2012 ARM as 6:49.60, set on October 6, 1973 at Thistledown (Cleveland, Ohio) by Eastern Promise, 6 years old and bearing 120 lbs.

Perhaps Flying Childers cheated. Perhaps he was really flying rather than running? In that case, Thoroughbred genetics has reversed rather than advanced in the past 291 years.

The Sequence of Time Adjustments

Selected time adjustments are now discussed based on the general boundary conditions in Figure 7.

Chapter Seven

As a *first* adjustment method, beyond using raw times, the times for each colt's *separate* six-furlong races were adjusted based on three levels for Man o' War and one for Secretariat.

Man o' War's adjustment included using: 1) the *average Speed Rating* for that day, equal to the Track Variant; 2) the *existing track record when Man o' War ran relative to Secretariat's track record* for the same distance—obtainable from the Speed Rating and 3) the *effect of the extra impost* on Man o' War, if any, including racing plate weight.

Secretariat's adjustment was relative to his Track Variant *only*. His time could not automatically be assumed equivalent to the current track record where he ran because that would negate Man o' War's adjustment to the same level. He also could not be credited with an impost adjustment since he always carried less weight than Man o' War.

Identical procedures are followed in the next chapter when adjustments are made to the finish times of both colts for the nine-furlong races.

If the adjusted data show apparent distortion from a normal distribution, simulation using random data selection *from* a normal distribution, as is required to resolve final time issues, would be meaningless.

Therefore, all adjusted data were checked for normalcy using the Shapiro-Wilk test. Since both Man o' War's and Secretariat's original data test normal using that test, only those adjustments maintaining the original normalcy were considered valid.

The Adjustment Process for the Six-Furlong Races

As an example of the method and to clarify interpretation, the process is now detailed for the first, and only the first, of Man o' War's six six-furlong races.

Man o' War & Secretariat

His final time for the July 5, 1919 race at Aqueduct, from Data List 10, was 73.00 s. His Speed Rating and Track Variant for that race were 90 and 12, respectively.

A further, but modest, assumption was made that 50% of the Track Variant value was attributable to the track speed for that day and the remaining 50% was due to the general running ability of the colts, to the jockey's ride or to some combination thereof.

This 50% Track Variant adjustment was applied to both Man o' War's and Secretariat's times. Thus, Man o' War was given time credit for one-half of his Track Variant value and for the *total difference* between his existing track record for each six-furlong race and the *average* of Secretariat's corresponding existing track records.

Since Secretariat could not logically be credited for time against the same track record serving as a baseline for Man o' War, the only remaining time credit to give him was for the assumed 50% effect of the Track Variant when he raced.

In support of the 50-50 Track Variant split, consider that the Travers times correlate -0.82 with the years it was run, from 1917 through 2008, and also correlate -0.64 with foal crop sizes for the same years. The large sample size, 91, lends cachet to the correlations.

These correlations are significant. Their squared values, or CODs, are 0.67 and 0.41, respectively, rounded to two places. The CODs imply that either 67 percent of changes in the 91 winning Travers running times can be explained, in some sense, by the progression of passing years—or, alternatively, that 41 percent of the running time changes are explained by related changes in the foal crop size. One has a choice in this regard.

Chapter Seven

The complementary differences between these values and 100 percent, 0.33 and 0.59, average nearly 50 percent, and these differences can be attributed to track effects.

Refer to the second section of Appendix H which explains how to test the significance of a single correlation value between data sets using the t-test.

Due to the liberal and often inconsistent fluctuations displayed by Thoroughbred data, it is thus reasonable to estimate that the 50-50 split attributed herein to Track Variant — with equal apportionment between track effect and ability, as earlier suggested, is justifiable.

Since the *actual* percent partition of the Track Variant cannot be accurately determined from the raw data alone, the 50/50 split is also the most equitable to give each colt. Appendix I offers further justification for using a 50/50 split vice other value combinations.

Continuing Man o' War's adjustment example, his Speed Rating of 90 indicates that his final time was 10/5 of one second (100 – 90 = 10), or 2.00 s, slower than the standing track record. Direct subtraction of 2.00 s from his time, 73.00 s, means that the *track record* when and where he ran was 71.00 s.

Secretariat ran just three six-furlong races as a juvenile compared to six for Man o' War. Therefore, *direct comparison on a race-by-race basis cannot be done* for their data. We must first *average* Secretariat's track records for six furlongs and then compare each of Man o' War's track records with that average.

This is equitable, especially since the mean is the most important and widely used measure of central location. (Ferguson 46)

Secretariat's average six-furlong track record (three races) was 69.00 s and Man o' War's track record for his first six-furlong race was 71.00 s. Therefore, Man o' War receives a *2.00 s adjustment* to this part of his original time.

Man o' War receives, as per the above explanation 1/2 of the time value for Track Variant effects. Since his track variant for the first six-furlong race was 12, this means that one-half of 12, or six, units of time are also subtracted from his original time. Recall that the basic time units for Track Variant and Speed Rating are *fifths of one second*, or 0.20 s. His Track Variant portion of the time adjustment is then 6/5 s or 1.20 s.

His remaining adjustment is for impost, including the weight of steel racing plates. It was 0.18 s for his first five six-furlong races and 0.13 s for his last such race. The difference of 0.05 s results from his final impost at six furlongs of 127 lbs. vice 130 lbs. for his first five such races.

Consequently, Man o' War's *total time adjustment* for his first six-furlong race was: 73.00 s – 2.00 s – 1.20 s – 0.18 s. The result is 69.62 s.

The same procedure was followed for each of his five remaining six-furlong races. The results for both Man o' War and Secretariat are in Data List 13. In the latter case, Secretariat is given an adjustment only for one-half of his Track Variant's value below par 100, as previously stated, for each of his six-furlong races. He receives no impost adjustment or adjustment of his track record to another standard.

Secretariat's first six-furlong race was at Aqueduct on July 15, 1972. He earned a Speed Rating of 90, with a Track Variant of 14. Thus, Secretariat's track record was 68.60 s.

Chapter Seven

This time is averaged with his remaining two track records— the average of the three then being subtracted from *each* of Man o' War's times. Secretariat receives one-half the time value of his Track Variant, namely, 7/5 or 1.40 s. Therefore, his original time, 70.60 s, becomes: 70.60 s − 1.40 s (7/5 s), or 69.20 s.

The same procedure applies to his remaining two six-furlong races. The results in Data List 13 imply times in seconds. List abbreviations are given in the following paragraph.

Data List 13

Man o' War Adjustments: 6 f						Secretariat Adjustments: 6 f			
$Time_0$	MTR	STRΔ	.5TV	Impost Δ	$Time_f$	$Time_0$	.5TV	STR_a	$Time_f$
73.00	71.00	2.00	1.20	0.18	69.62	70.60	1.40	69.00	69.20
72.40	70.40	1.40	1.30	0.18	69.52	70.80	1.30	69.00	69.50
71.27	70.20	1.20	0.90	0.18	68.99	70.00	1.40	69.00	68.60
72.00	70.40	1.40	0.80	0.18	69.62				
73.00	70.40	1.40	1.40	0.18	70.02				
71.60	68.60	0.00 *	2.10	0.13	69.37				

* Man o' War's TR (68.60) was *less than* Secretariat's average TR (69.00). Thus, no subtraction was possible and zero time credit resulted.

Abbreviations, read from left to right, in the data list are: $Time_0$ = original, unadjusted time; MTR = track record where Man o' War raced, determined from his separate speed ratings; STRΔ = MTR − STR_a (Secretariat's average track record); .5TV = one-half the time value of the Track Variant that is given to each colt; Impost Δ = time value assigned Man o' War for higher imposts; $Time_f$ = final adjusted time for each colt.

Since the assumption is that overall Thoroughbred species running ability does not change significantly in 53 years (generally millions of years are required for real genetic change (Meyer 15), and evidence presented earlier in this chapter supports a uniformity conclusion), an actual direct comparison of *pure track differences* across the 53-year gap, rather than differences in running ability, should be reasonably approximated by subtracting Secretariat's average track record from each of Man o' War's separate track records.

The above calculations set Man o' War's cumulative *average adjusted* six-furlong time at 69.52 s with a standard deviation of 0.34 s. Secretariat's final *average adjusted* six-furlong time becomes 69.10 s with a standard deviation of 0.46 s.

These adjustments are equivalent to an average of 2.70 s being deducted from each of Man o' War's six-furlong times and to an average of 1.37 s being deducted from each of Secretariat's. The resultant 1000 random simulations based on these adjustments show Man o' War receiving 235 faster times out of 1000 and Secretariat receiving the remaining 765.

The equivalent sequence of ten sets of 100 simulations each assigned the following faster times to Man o' War: 26, 19, 21, 22, 24, 21, 23, 24, 26 and 29.

One must realize that this simulation is equivalent to what would be generated (within general rounding error) if the difference between the adjustments, 2.70 s—1.37 s = 1.33 s were *directly subtracted* from each of Man o' War's six-furlong times with *no adjustment* given to Secretariat. A random sample of 1000 using this latter procedure show Man o' War receiving 285 faster times of 1000 compared to Secretariat. This is basically within the fluctuation range for the previous value of 235 and is expected for this number of random samples.

The original and the adjusted data for both colts are normal by the Shapiro-Wilk test. The p-values of the adjusted times are 0.6350 and 0.6369 for Man o' War and Secretariat, respectively.

Any result greater than 0.05 for the Shapiro-Wilk test denotes that the sample could have been randomly selected from a normal distribution with the stated p-value.

The foregoing modifications having been stated and applied, under which the colts' running times were altered, leave a difference of only 0.42 s between Man o' War's and Secretariat's average six-furlong times. Thus, the difference is: 69.52 s − 69.10 s = 0.42 s.

Also note that a t-test of the adjusted times gives the non-significant value 0.092. This places the resultant adjustment time difference, 1.33 s, at t = 0.197 in Figure 7. The placement is not explicitly shown in the figure to maintain simplicity. However, its equivalent mean value, 70.88 s, assigned to Man o' War for that adjustment, results in a logical placement on the line, between 70.71 s and 71.34 s.

A general explanation of Man o' War's impost adjustment is now given and includes the amount by which steel (versus aluminum) plates might affect running times. The effect is actually small.

Explanation of the Impost Adjust Procedure

The question is now fully addressed concerning what is an unbiased way of adjusting Man o' War's six-furlong times for his impost relative to Secretariat's for similar races.

For his six six-furlong races, Man o' War carried, in chronological order by racing date, 130 lbs. for the first five races and 127 lbs. for the final race. Adding the *calculated* weight of steel plates he wore, 2.62 lbs., increases the former values to 132.62 lbs. and to 129.62 lbs., respectively.

To determine plate weight a set of aluminum racing plates was procured from a well-known supplier. A single plate was weighed and registered 3.6 oz, or 0.225 lbs. Therefore, multiplying by four gives the total weight of 0.90 lbs. to the nearest 0.01 lb. for the complete set.

It is straightforward to determine the weight of an identical steel plate simply by referencing a manual of materials engineering and determining that a typical alloy steel from which plates can be made weighs 2.91 times as much per gram as aluminum. When the required calculations are done, the weight of a complete set of four such plates equals 2.62 lbs.

This weight is likely high because steel plates could be thinner than aluminum and still be able to resist the distorting forces affecting them during running. The present adjustment considered them equal in all dimensions, thickness included.

It is possible that they are no more than three-quarters as thick as the corresponding sized aluminum shoe. They would then be distinctly lighter, at slightly below 2.0 pounds. However, in the interest of absolute fairness to Man o' War, and not having exact dimensions for his plates, the above weight value stands and is used in the actual adjustment procedure.

In GHA, estimates of impost affect on running times were referenced to the formula for kinetic energy from classical physics. (Justice 26) While the suggestion was correct from a theoretical standpoint, it apparently does not apply well to reality, as reflected in published data. (ARM 911)

Chapter Seven

The 2012 ARM lists the fastest times for dirt at numerous North American tracks for the years 1991 through 2011 at distances of six, seven, eight, eight and one-half, nine and ten furlongs.

Data List 14 summarizes the main characteristics of these data regarding age, impost and time ranges. The sample sizes are all large, or significant, ranging from 48 to 54. The data span 20 years in each case and represent 312 horses. In the list, ages are in years, imposts are in pounds and final times are in seconds.

Data List 14

Distance	Age Range	Impost Range	Time Range
6f	2 – 7	113 – 124	66.49 – 67.54
7f	2 – 7	111 – 123	79.70 – 80.69
8f	3 – 7	107 – 126	92.24 – 93.83
8.5f	3 – 7	110 – 125	99.18 – 100.46
9f	3 – 6	111 – 126	105.35 – 106.69
10f	3 – 6	110 – 126	118.33 – 120.45

Appendix J presents charts for the complete data sets at each of the above distances. The trend equation and its COD for each distance are also given directly below the charts, as is the sample size.

The principal points to note are that none of the trends of running time versus impost show a higher correlation than that for nine furlongs, -0.290, and it denotes not much better than a random relationship. It also indicates a *negative* trend—contrary to expectation.

In fact, four of the six trend lines, including the nine-furlong case, have negative slopes. This is reflected in their negative correlation values and also by the downward slant of the trend lines from left to right. The implication is that, for these cases, the *greater the impost* a given horse carried, the *faster* was its running time. This contradicts common sense, but it represents fact in these cases.

Therefore, *it is not necessarily true that higher imposts negatively affect a horse's running ability*, since they did not for these horses running at or near world record times. Inconclusive and inconsistent correlations between impost and running time were noted in GHA and BG. (Justice 44; 38). See Davidowitz, for example, on this point. (122)

Despite these inconsistencies, adjustments were made for Man o' War's six-furlong races based on the data. The adjustment method and its rationale are now detailed.

Using Confidence Intervals to Refine the Impost Adjustment

Using confidence intervals is one way to attempt compensating for the generally small correlations between impost and time. With such small correlations, one cannot be certain that the predicted time value for a given impost will be remotely close to the *predicted average for that impost*, as estimated by the regression equation.

In fact, for six furlongs, the correlation is just 0.255 and the corresponding COD is 0.065. This means that changes in impost account only for 6.5% of change in running times.

Restating this, one may not be within 7% of the actual average value by which Man o' War's time was affected for a given impost using the available regression data.

Therefore, it was decided that an equitable way to find a reasonably accurate estimate was to find the standard confidence interval around the predicted average. Confidence intervals are easily calculated from appropriate z-values multiplied by the standard error of estimate.

Confidence intervals may be calculated either manually, with the help of a table of z-values, or also by using Excel's CONFIDENCE function from the statistical menu. Since manual calculation is more informative, it is now illustrated.

The general formula for manual calculation is:

$$CI = \bar{x} \pm z_{ci} \cdot s_{\bar{x}}$$

In the formula, CI indicates confidence interval, $\bar{x}$ is the mean of a data sample, z_{ci} is the z value for the particular confidence interval (ci) used, e.g., 95%, and $s_{\bar{x}}$ is the sample standard error of estimate.

To perform the calculation, multiply the z_{ci} value appropriate to the preferred confidence interval by the standard error of estimate, $s_{\bar{x}}$, found directly from Excel's Linear Estimate function output table. As discussed earlier, the standard error of estimate is the third item down the second or rightmost column of the LINEST output table. It is also shown as item 6 in Appendix D.

The three most commonly used confidence intervals, 90%, 95% and 99%, have respective z_{ci} values: ± 1.64, ± 1.96 and ± 2.58.

The standard error of estimate, $s_{\bar{x}}$, is actually the *standard deviation of the mean value obtained* when repeated samples are taken from a normally distributed population, e.g., by taking ten random samples of 5 items each and determining their individual means and then finding the standard deviation of those ten means.

In the present case of estimating the impost effect at six furlongs on Man o' War by using the appropriate chart from Appendix J, the complete linear regression table, not shown in the Appendix, gives the standard error of estimate as 0.227737.

This value was rounded to two decimal places, 0.23, for quicker calculation because the regression is only accurate to approximately 6.5%, and no significant inaccuracy occurs.

The *mean* impost correction predicted by the six-furlong linear regression for Man o' War—carrying 130 lbs. to Secretariat's average 117.33 lbs.—is 0.25 s.

However, a 95% confidence interval around this mean value indicates that the interval's lower boundary is located at -0.0659 seconds below the mean. Rounding this value to -0.07 and subtracting it from the mean gives the value 0.18 s—the actual time credit given Man o' War for his first five six-furlong races.

This value was deemed equitable, given the low prediction level of the six-furlong world-record linear regression. It was considered better to err on the side of caution than otherwise.

However, the value 0.18 is still well within the first standard deviation below the mean—which actually extends to -0.23 s below the mean - and is a reasonable concession to error without unduly penalizing either colt.

For the initial five six-furlong races, the impost-based time adjustment for Man o' War was, therefore, 0.18 s. Since Secretariat carried less weight than Man o' War for all races, he received no impost adjustment.

During the final six-furlong race, Man o' War carried 127 lbs. Reasoning analogous to that for the previous five six-furlong races determines that 0.13 s is an equitable time adjustment for the final race.

When the final adjusted times for both colts were averaged and compared, Man o' War's average time adjustment for his six-furlong races was 2.70 s with a standard deviation of 0.43 s.

Secretariat's average time adjustment for these races was 1.37 s with a standard deviation of 0.06 s. Thus, the relative difference between the two adjustment times was 1.33 s.

A t-test between the two sets of final *adjusted times* had F value 0.54. This implies that the variance between the data sets was *not significantly different* since the F value exceeded 0.05.

Given that the variances tested homogeneous, a *one-tailed t-test* was used. The result was $\alpha = 0.039$. This value is below 0.05. It therefore implies that a *significant difference exists* between the two sets of adjusted times. However, it is a borderline significance level considering the small sample size. The adjustment seems, therefore, tentatively reasonable.

Based on this initial analysis, a fair adjustment between the data for the two colts has been achieved. Therefore, no further data adjustments were needed, other than for possible explorative value, and no special era effects need be invoked in premonition that unfair comparisons will follow.

All such premonitions are basically mere chimeras—psychological will-o'-the wisps.

It is striking that the average adjustment value of 2.70 s for Man o' War is of the order previously referenced for Farley. (337) It is definitely not unreasonable.

A Second Type of Confidence Interval

The confidence interval just calculated applies when a sample or series of samples are taken from a population in which the mean and standard deviation may or may not be known.

The formula, as stated, is handy because it is easily applied and because its parameters are readily available. Necessary z-values are found in tables in nearly all beginning statistics texts. The population mean and standard deviation, if unknown, can usually be reasonably approximated by the sample mean and standard deviation as obtained from that population. This is standard procedure in statistical analysis.

In other applications, however, a related but different confidence interval must be calculated. That interval is required in cases of linear regression where it is desired to calculate a *single future value* from the regression formula. (Walpole 332)

If, for example, you wish to predict the outcome of a horse race in which you have gathered data for performing linear regressions for several horses, it is advisable to calculate separate confidence intervals for each colt at least at the 95% confidence level.

Based on the standard deviations obtained from these confidence intervals, random-number simulations (Excel RNG) can be performed for each horse and the probability determined of whether a favorite steed would likely win the most events out of 100.

The outcome then dictates whether betting should or should not proceed.

The random number generator requires both a *mean and standard deviation* before it can produce random numbers, hence the need for determining a confidence interval.

The appropriate standard deviation can be calculated once the confidence interval is determined by, for example, dividing one-half the magnitude of a 95% confidence interval by 1.96, which represents its equivalent number of standard deviations.

It is tempting to use the standard error of the mean automatically produced by Excel's LINEST routine (item #6 in the Appendix D table) for the standard deviation, but this is not the true estimate needed for actual confidence intervals.

It was used as a shortcut in GHA where its accuracy was sufficient to illustrate a point. It should not, however, be used as the basis for betting.

Appendix K shows that the standard error of estimate, $s_{\bar{x}}$, is just one of *three* factors needed to calculate a regression confidence-interval component, A. The other factors are fully explained in the Appendix.

The general form for applying the calculated confidence interval is, analogous to the previous equation for a single-sample interval:

$$CI = \hat{y}_0 \pm A.$$

In the formula, $\hat{y}_0$ is the average value of the dependent variable *predicted by* the regression equation for a *single known value* of the independent variable, x_0, and A is the actual confidence interval.

Appendix K also describes how A is calculated.

Linear Regression and the Trapezoid Adjustment Method

A second time adjustment method, using linear regression with a "trapezoidal" approximation, is now discussed for Man o' War.

This technique is partly exploratory since, to my knowledge, it has not been previously used.

An Internet posting gave me the idea of calculating a trapezoidal area (Cardenas 2), although it did not specifically use this technique, beneath linear regressions—such as for comparing Man o' War's and Secretariat's records to the current world-record.

If the method proves viable, it bodes well for unbiased comparisons because it provides an additional perspective on the data. If it proves impractical, no real harm is done and the method is not further applied.

The following is a general explanation and investigation of this approach.

Linear regression can be used as an ancillary method for investigating data adjustments and comparisons over a *broader range* than was just done using the adjustment method relying on single track record comparisons.

Rather than using only the six-furlong distances, linear regression allows looking at *all distances* each colt ran in a given racing year and then comparing that year's data as a whole. Only the juvenile year is discussed now.

The section of this chapter entitled 'Basic Juvenile Year Data' gives a complete list of juvenile distances and times for both Man o' War and Secretariat. Refer to that section and to Appendix A for reference questions concerning the raw data.

Since familiarity is already established for using Excel's LINEST routine, only the results of that application are now given.

The equations derived from Excel's LINEST routine for Man o' War's and Secretariat's juvenile times are, respectively, 95.605 x + 0.488 and 103.438 x − 7.088.

Chapter Seven

An exact plot of these equations is not given. The regression lines fall too close together when plotted on a single graph having equal scales for both equations.

Refer now to Appendix L. Generic plots of each equation, compared with the 2011 world dirt records, are presented there. For greater clarity, only *relative* horizontal and vertical data point spacing is used. The numbers on each graph referring to specific times are, however, accurate and were taken from the respective linear regressions.

These equations and plots are used to obtain a *single numerical value* for simultaneously comparing *all* juvenile races between these two runners, rather than comparing them for separate races only.

It happens that a straightforward *adjustment of the equations* relative to the world-record trend line can be made. The technique is explained in Appendix L.

Referencing to the world-record trend line for distances compatible with those the colts raced provides an *absolute reference level* analogous to that used in scientific work. Such a level is sorely needed in the Thoroughbred racing industry.

The trapezoidal adjustment requires 1) *finding the area* beneath each colt's linear trend line included *between* the desired comparison distances and 2) finding the area beneath the corresponding world-record line between the same end points (distances run, or abscissa values) along the horizontal axis.

These areas are trapezoidal, as can be seen from graphs L.1 and L.2. Therefore, the standard area formula: ½ x height x (sum of the bases) is used to determine them. In these examples, the trapezoids are oriented so that their '*heights*' lie along the *horizontal axis* and their '*bases*' are *parallel* to the *vertical* axis.

The heights are the lengths of the line segments AD. The bases are the lengths of the line segments AB and CD.

The trapezoidal areas in each graph are labeled ABCD. After the separate areas are determined, each colt's area is then *reduced*, by lowering its trend equation line, until its area *equals* the area under the world-record line between the same end points.

This technique preserves the proportional spacing between all data points included within any given regression equation. It **may**, therefore, provide an overall more accurate assessment for each race and for the combined races since the regression lines have a high COD with respect to their data points. Some error, however, always exists irrespective of technique.

The amount *each* regression line must be *lowered* so that Man o' War's and Secretariat's regression lines are *equally proportioned to the world-record* line represents the *amount by which each* of the colts' run times must be adjusted so that his data are mutually compatible with any track differences existing in the fifty-three years separating them.

For *example only*, using simple numbers, but not actual racing times—if Man o' War's juvenile linear regression must be lowered by 3.95 s *relative to the world-record* trend and Secretariat's juvenile linear regression must be lowered 1.27 s. relative to the *same standard*, then the *difference* between those values, namely, 3.95 s – 1.27 s = 2.68 s, is the number of seconds *each* of Man o' War's times must be *reduced* relative to Secretariat's so that his times can compare fairly with Secretariat's. That, at any rate, is the theory.

Since Appendix L gives a detailed example calculation for both Man o' War's and Secretariat's juvenile races, only summary results are given now for both the juvenile and sophomore racing seasons. The reader is encouraged to verify the results for better understanding. The numbers in Data List 15 give the **areas units only** for each colt, the world-record area units, and the final adjustments, Delta, in seconds, needed for Man o' War relative to Secretariat.

Data List 15

	Juvenile Year				Sophomore Year		
Man o' War	Secretariat	World	Delta	Man o' War	Secretariat	World	Delta
72.20		67.02	5.18	110.58		104.40	6.18
	70.49	67.02	3.47		107.55	104.40	3.15
		Final:	-1.71			Final:	-3.03

The final time adjustment value, -1.71 s for the *juvenile* year is found by *direct subtraction* of 3.47 from 5.18, just as the value -3.03 s for the sophomore year is found by direct subtraction of 3.15 from 6.18. This is a convenient calculation due to the selection of an equivalent trapezoidal height of *one unit* along the horizontal axis.

For the juvenile year the horizontal range is from 5.5 f to 6.5 f (1 unit) ; for the sophomore season the horizontal range is from 8.5 f to 9.5 f (1 unit).

The original finish times and the adjusted times for the six-furlong races are given in Data List 16. The means, standard deviations, F test and t- test results, plus the results for normalcy check using the Shapiro-Wilk test are also included. Times are in seconds.

Data List 16

	Man o' War			Secretariat		
	Original Time	Adjust1	Adjust2	Original Time	Adjust1	Adjust2
	73.00	69.62	67.82	70.60	69.20	67.13
	72.40	69.52	67.22	70.80	69.50	67.33
	71.27	68.99	66.09	70.00	68.60	66.53*
	72.00	69.62	66.82			
	73.00	70.02	67.82			
	71.60	69.37	66.42			
Mean:	72.21	69.52	67.03	70.47	69.10	66.99
SD:	0.72	0.34	0.72	0.42	0.46	0.42
F test:	0.54	0.51	0.54	0.54	0.51	0.54
t-test:	0.003	0.08	0.47	0.003	0.08	0.47
S-W:	0.69	0.64	0.69	0.46	0.64	0.46

*The list values for Adjust 2 for both colts were based on subtracting 5.18 s from each of Man o' War's times and subtracting 3.47 s from each of Secretariat's times. The same result, within rounding error, would follow if 1.71 s, the difference between the original subtracted quantities, were directly used for Man o' War's time adjustment only. Adjust 1 refers to using Track Variant and track records.

Since all F test, t-test and Shapiro-Wilk (S-W) values are above 0.05, except for the highlighted t-test value 0.003 between Man o' War's original set of six-furlong times and Secretariat's original times, no significant difference exists between these data.

The Shapiro-Wilk values also indicate that all samples, from original through the adjustments, could derive from normally distributed populations.

There is, therefore, no indication that the original data were distorted by the time adjustments for either colt.

Note, particularly, that the variation between the resultant trapezoidal adjustments for the juvenile and sophomore years,

(3.03 s − 1.71 s = 1.32 s) is not inconsistent with variables such as: 1) difference in the CODs of the juvenile and sophomore regression equations, 2) increased maturity of the colts, 3) differences in the tracks run and 4) variations from one year to the next in track surfaces, even for identical tracks.

The relative closeness of these two values, considering data fluctuations, presents a favorable argument that the *true* difference in track speeds between Man o' War's races and Secretariat's was near two seconds for eight furlongs, consonant with Farley's suggestion. (337)

Although this chapter was basically devoted to the juvenile year, it seemed permissible and more informative to introduce the trapezoidal results for the sophomore year also. The column labeled 'Adjust 2' in the preceding list refers, however, only to the trapezoidal method applied to the juvenile races.

The following chapter discusses the simulation results from these adjustments and the concept of applying z-scores to the data of both colts. The z-score virtually guarantees that no era effects, even *if they did exist*, are involved in its calculation since it compares data among peers only.

Absolute, true, and mathematical time, in and of itself and of its own nature, without influence from anything external, flows uniformly, and by another name is called duration.
~ Sir Isaac Newton

Chapter Eight

Handicapping Man o' War and Secretariat as Juveniles

The term 'handicapping' is not strictly associated with simulating races via computer, but the results of both are comparable. When Excel's Random Number Generator (RNG) function is used to compare how Man o' War and Secretariat (any two horses, for that matter) would fare if they competed in multiple races at the same distance, the results are expressed as a percent of wins for each. This is merely a different expression for the 'odds' that are placed on horses by race handicappers.

The RNG uses the parameters *mean* and *standard deviation* of each colt and 'draws' whatever number of desired times it is assigned to produce from a *theoretical normal distribution* having the same parameter values.

The value of each generated time in the simulation changes according to the RNG's internal algorithm. Considering this, readers desiring to duplicate such values should understand that *stated values* in Data Lists involving sequences of simulated numbers will differ somewhat for each simulation. Thus, these values do not contract each other. They merely reflect the nature of the process.

This chapter compares Man o' War and Secretariat at six furlongs—the sole common distance they ran multiple times as juveniles.

Four simulations were used to determine how selected time adjustments among those discussed in Chapter 7 affected the number of expected faster times Man o' War gained compared with Secretariat at six furlongs.

The first simulation used only the raw scores of both colts—the recorded, unmodified times for six furlongs published in the two Champions editions by Daily Racing Form. (28; 279)

The second simulation was based on the adjustments to Man o' War's and Secretariat's times for six furlongs based on the Track Variant and track record criteria, labeled 'Adjustment 1' in Chapter 7, while a third simulation was based on the trapezoidal method, Adjustment 2.

The fourth simulation was based on a final adjustment using z-scores to relate Man o' War's and Secretariat's *original* data and the forty-eight fastest six-furlong results from the years 1991 through 2011. (ARM 911) The means and standard deviations of Man o' War, Secretariat and the six-furlong world records were compared using ten separate simulations of 100 trials each. Excel's RNG was again used based on random drawings from a normal distribution.

Data List 17 presents the four simulation results in terms of how many faster times Man o' War's record indicates he gained using each adjustment method. In other words, the '3/100' given for the first trial result under 'Raw Data' means that Man o' War showed three faster times in the initial sequence of 100 simulations.

For the final adjustment method, using z-scores combined with the average world record for six furlongs, Man o' War received 504 of 1000 fastest times. Therefore, Secretariat automatically received the remaining 496. The closeness of this

Chapter Eight

result is not surprising because subtracting 5.18 s from each of Man o' War's times and 3.47 s from each of Secretariat's raw times (or, equivalently, subtracting 1.71 s from each of Man o' War's times while leaving Secretariat's times unadjusted) results in both colts having identical means, within rounding error, and their standard deviations differing by 0.30 s. Man o' War's SD was 0.72 and Secretariat's was 0.42.

To emphasize that RNG results *differ from trial to trial*, all ten results (100 trials each) are included in Data List 17 for each of the four comparison methods. The nature of random sampling is thus emphasized. Abbreviations used are: TV = Track Variant; TR = Track Record; WR_a = World Record average for six furlongs.

Data List 17

Raw Data	TV/TR Method	Trapezoid Method	z-Score/WR_a
3/100	37/100	52/100	47/100
2/100	31/100	55/100	57/100
2/100	27/100	49/100	53/100
2/100	37/100	41/100	51/100
0/100	32/100	50/100	57/100
1/100	29/100	46/100	51/100
2/100	24/100	45/100	35/100
3/100	31/100	58/100	45/100
3/100	24/100	53/100	53/100
2/100	32/100	57/100	55/100
Totals: 20/1000	304/1000	506/1000	504/1000

Note the nearly identical results between the Trapezoidal and z-score adjustments.

Since the Trapezoidal adjustment technique was admittedly exploratory, the results lend strong support to it because z-scores are statistically well established.

A Shapiro-Wilk normalcy test for each of the above sequences gives the following probabilities, corresponding to the left-to-right data columns above, that the sequences are from normal populations, with p-values: 0.0398, 0.5654 0.9706 and 0.0714. These values can change, however, with each new simulation. That is the nature of randomness.

Only the first data column, representing the original raw data, tests non-normal and that value, 0.0398, is marginal, $\alpha = 0.05$ being the significance cutoff value. It probably results from the multiple occurrences of the values '2' and '3.' The Shapiro-Wilk test regularly gives a non-normal result for small samples having repeated values.

This is an instance wherein common sense must prevail in the interpretation. Since all the other sequences were judged within the normal range, it makes sense that the original data also represents a normally distributed population.

Note that the *first adjustment* method, TV/TR in column two, results in a difference of 2.70 s between Man o' War's original, unadjusted average time for six furlongs and his final average adjusted time for the same distance: (72.21 s - 69.51 s = 2.70 s).

The second adjustment method (Trapezoidal), relative to the world record sample, gives a difference of 5.18 s between Man o' War's original average time for six furlongs and his adjusted average time: (72.21 s – 67.03 s = 5.18 s).

Chapter Eight

The details of how the world-record adjustment was made are given later in this chapter when z-scores are discussed.

These adjustments are greater than Farley's estimate (almost 2 seconds faster to the *mile*) cited previously. (337). If a near proportionality is assumed, for ease of calculation, between distance and Farley's value, then six furlongs would merit 1.50 s adjustment. The above suggested trapezoidal/world-record adjustment is nearly three and one half times greater.

It is important, however, that three of the above four adjusted data columns have normal distributions under the Shapiro-Wilk test and that the Raw Data is *marginally* non-normal. This indicates that, in fact, data distortion was not introduced by the adjustments.

Therefore, nothing is suspect about these results other than a question of their ultimate accuracy—a question that always arises when random sampling is used.

As already noted, the values in the preceding list were generated using Excel's Random Number Generator (RNG) function. For those new to simulation using this application, the following guidelines are presented:

1. To perform a basic simulation for *two sets* of times, label the top cell in each of three *adjacent* Excel spreadsheet columns (e.g., A1 through C1) 'Man o' War' (use preferred abbreviations for all labels), 'Secretariat' and 'Compare.'

2. Select the topmost available cell, A2, under the 'Man o' War' label in cell A1.

171

3. Click 'Tools,' then 'Data Analysis,' then 'OK Random Number Generator' choices.

4. In the miraculously appearing Random Number Generator (RNG) Dialog Box enter, consecutively, '1,' '100,' 'Normal,' and Man o' War's mean and standard deviation values for his run times in the provided fields.

5. Select the 'Output Range' button, then enter the cell range 'A2:A101.' Click 'OK.' This automatically generates 100 *random values* of run times in cells A2 through A101. The RNG application generates these values from a theoretical normal distribution having the specified means and standard deviations.

6. Repeat steps 4 and 5 for cells B2 through B101 to generate Secretariat's corresponding 100 random comparison values.

7. In cell C2 enter '=if(A2<B2,1,0)'. When the ENTER key is pressed either a '1' or a '0' is then automatically entered by Excel in cell C2, depending on whether or not the stated 'if' condition is met. In this case, '1' signifies that Man o' War's first random time in cell A2 was *less than* Secretariat's random time in cell B2. A '0' signifies the opposite condition. Next, copy and drag the results in cell C2 to fill the consecutive cells down to and including C101 with either '1s' or '0s.'

8. Select cell C102. Click the function button marked 'Σ' in cell C102. The entire column of cells should now have a 'sparkling' border which Excel calls the 'marquee.' Now press ENTER. The Σ function automatically sums the '1s' in column C, thus showing directly how many simulated races of 100 showed Man o' War having faster times than Secretariat.

9. Repeat the entire sequence of steps 1 – 8 above to run additional simulations, as desired. Virtually any number of random values can be generated by excel limited only by the available rows provided by the application's design.

There are slightly more than one million such available rows in Excel's 2007 version. (Harvey 229)

The next section discusses z-scores related to the juvenile years of our protagonist colts. It also explains how the world-record data in the forgoing list were derived using the z-scores.

Freedom from Era Effect Guaranteed using z-Scores

The z-score is perhaps the easiest statistic to explain and to apply. It was used in the book BG to rank Man o' War and Secretariat on a tentative basis relative to their peers. It is *absolutely independent* of any era effect that *may or may not* exist since it gives only relative rankings of horses running during similar years as the subject horses.

The z-score is also highly convenient to interpret. Its value *exactly represents* by how many standard deviations any given value, or time for horses, is either above or below the mean of the sample. Negative z-scores indicate values below the mean—faster times.

In BG, race results for major races seven or eight years on either side of the years in which Man o' War and Secretariat ran were compared with their times for the same races.

Only the result is given here since this chapter approaches z-scores from an entirely different and more intuitively meaningful way than the BG method.

In the BG method, data from seven of Man o' War's sophomore races and comparable data from nine of Secretariat's sophomore races were compared.

The result was that Man o' War's average z-score was -0.918 for N = 7 and Secretariat's average z-score was -1.092 for N = 9. A t-test run on the complete sets of z-scores was not significant. Its value, $\alpha = 0.7944$, supported the conclusion that the two sets could likely have come from a common population or, alternatively, from two populations having equal means.

By definition, a z-score is simply the *difference* between a given colt's time for a certain distance and the *mean value* for *all* the colts running the same distance—that difference then being *divided by* the *standard deviation* for all times combined.

Man o' War's average z-score, -0.918, means that his times for the seven races averaged lower than 82.07 percent of all colts running the same set of races in years comparable to his—nearly one standard deviation below the mean.

Secretariat's average z-score, -1.092, means that his times for nine races averaged lower than 86.26 percent of all colts running the same set of races in years comparable to his—slightly more than one standard deviation below the mean.

A Shapiro-Wilk test for normalcy, run for both sets of z-scores, gave probability values of: 0.052 and 0.527 for Man o' War and Secretariat, respectively. Therefore, both sets of scores can be considered as originating from normally distributed populations. This implies that no data distortion resulted from directly comparing the two sets of values.

Chapter Eight

These results, therefore, suggest that both colts were essentially equal, with respect to their individual peers, for the times in which they ran.

We now wish to test this conclusion by applying the z-score differently.

Using Specific Race Results to Derive z-Scores

Rather than compare Man o' War and Secretariat for races spread some seven or eight years on either side of their racing seasons, z-scores can also be derived for *each specific race* each colt ran. This will be done for both the juvenile and sophomore years, with only the juvenile results given in this chapter.

The method is as follows:

Daily Racing Form publishes the winning time and the lengths by which the winner won. It also shows the lengths by which the place, show and—depending on the field size—the first of the remaining field horses (fourth-place and lower horses) were behind when the winner finished.

These 'margins' or length numbers are *superscripted* immediately after the name of each respective finisher. A typical Daily Racing Form running line looks like this: MrChubs126^{5}SuperColt128^{4}CaptKirk117^3.

This line is interpreted: MrChubs, carrying 126 lbs., won by five lengths over SuperColt carrying 128 lbs. SuperColt was second, four lengths ahead of CaptKirk in third and carrying 117 lbs. CaptKirk was three lengths ahead of the fourth horse—the first of the field horses, assuming the chart shows four or more entries in the field.

If only three horses were entered, there would be no superscript '3' following CaptKirk's impost value.

The first field horse doesn't rate getting his name mentioned. Perhaps he was MrSpock?

Caution must be exercised when interpreting the past performance chart. Remember that the finish time given after the final call point is the *winner's* finish time, and *not the time* of any other horse in that race—*unless that horse won.*

One example of performing this technique, using Secretariat's data, is now presented. The remaining z-scores are only listed. The reader is urged to check them against the Daily Racing Form past performance charts [for instance, from the Champions 1999 or 2005 publication (DRF 279)] to gain experience in handling this type of data.

The final race of Secretariat's juvenile year was the Garden State Stakes held on November 18, 1972 at Garden State Park in New Jersey. It was an 8.5-furlong race which Secretariat won by 3.5 lengths in 1:44.40 over five other horses. His combined Speed Rating-Track Variant for the race is listed as 83-23.

Since the past performance chart gives the *lengths* which the *place, show* and, sometimes, the first *field* horse finished behind the winner, their finish times can be calculated.

If the horse that 'showed' was, numerically, the last horse entered in the race then, obviously, there are no additional lengths to calculate by which that horse finished ahead of the field—because there was technically no 'field' in that race.

Nonetheless, the Daily Racing Form format gives a ready-made sample size of at least 3 or 4 for most races—unless the

Chapter Eight

race was a 'walkover' or only two entrants were involved—as in a match race. Those are, however, relatively rare events.

None of the times are given for any horses which ran the previous races for that day at that same track and distance—unless one has their charts—but those data are not needed for calculating the z scores for a particular horse.

Secretariat's winning time of 1:44.40 is better expressed as 104.40 s in decimal format. This is by far the preferred form for doing calculations.

For this race, the place, show and first field horse are listed as being, respectively, 3 ½, ½ and ¾ lengths, in that order, behind Secretariat when he passed the wire. Recall that this notation means that the *place horse* was 3 ½ lengths behind Secretariat, the *show horse* was then *another* ½ length behind the *place horse*, and the first of the *field horses* was *another* ¾ length behind the *show* horse, as per the earlier example.

Therefore, the *cumulative lengths* behind Secretariat were: 3 ½, 4, and 4 ¾.

Daily Racing Form equates each length to 1/5 second, or 0.20 s. However, this is not remotely accurate from a physical standpoint. For instance, Secretariat averaged 12 s per furlong in winning the Belmont by 31 lengths.

Since there are 660 feet per furlong, Secretariat's average speed was 660 ft ÷12 s or 55 ft/s. One length is probably closest to eight feet, of the *typical estimates* assigned to one length by horseplayers. It is intended to represent the distance from a typical Thoroughbred's nose to the furthest rearward portion of his rump.

Assuming, for uniformity of discussion, one length is exactly eight ft, the speed of 55 ft/s translates to nearly 6.9 lengths per second. Since there are five one-fifth parts in one second, then 6.9 lengths equals 6.9 ÷ 5, or nearly 1.40 lengths per *one-fifth* of one second for Secretariat running the Belmont—not 1.00 length.

It is, therefore, easily seen that the time-honored equating of one length of distance to one-fifth of one second of time is inaccurate. Expert handicappers have long realized this fact. (Davidowitz 108) Tradition, however, is tradition.

To accurately calculate the finishing times of the horses lagging the winner, start by multiplying the number of lagged lengths by eight. This equates one length to eight feet and is probably more accurate than most estimates. If this length is used consistently for all runners, absolute accuracy is not an issue.

Using only the place horse for this example, 3 ½ lengths behind translates to 28 feet that he finished *behind* Secretariat.

Subtract 28 from the *nominal race distance*. Since the Garden State Stakes was 1- 1/16 miles, or nominal distance 5610 feet, subtracting 28 feet gives 5582 ft.

Therefore, the place horse had run 5582 ft to Secretariat's 5610 ft at the race's finish. It took him the same time to run that distance as it did for Secretariat to run the entire nominal distance, 5610 feet. Hopefully this is relatively obvious.

Divide 5582 ft by Secretariat's final time, 104.4 s because that's the time it took the place horse to reach *his* given distance, 28 ft behind Secretariat at the actual finish.

Chapter Eight

The rounded result is 53.47 ft/s. This is the place horse's *average* speed for the race *to that point*. Now assume that he maintains this speed for the remaining 28 feet to the finish line. This is a reasonable assumption for generally short distances and because the non-winning horses are almost always trying to close on the winner.

Take the reciprocal of 53.47 ft/s (1/53.47 s/ft) and multiply it by 28 ft. The result is 0.52 s (you should see that the 'ft' portions cancel during the multiplication) rounded. Add this extra time to Secretariat's winning time of 104.40 s to find the place horse's finishing time: 104.40 s + 0.52 s = 104.92 s.

Perform the same calculations for the show horse *and* the first field horse, as appropriate. In this case their respective finishing times are: 105.00 s and 105.11 s. Now find the *average and standard deviation* of the set of four finishing times, *including* Secretariat's.

The results are: $\bar{x}$ = 104.86 s and S.D. = 0.31 s. Statisticians consider the sample mean, $\bar{x}$ to be the best estimate of the population mean. Therefore, 104.86 s is the best estimate, using this race as *one sample* of size four from the *population* of equal-distance races held that day, of the average for all the day's eight and one-half furlong races.

Secretariat's z-score, therefore, is determined by subtracting the sample mean just calculated *from* his winning time and *dividing* the result by 0.31, the sample standard deviation, since it is also the best estimate of the population standard deviation. Thus, Secretariat's z-score for the Garden State race was: (104.40 − 104.86) ÷ 0.31 = -0.46 ÷ 0.31 = - 1.484 to three decimal places.

After the dust from the juvenile races and their z-score calculations settles, the results are as given in Data List 18, with the year's final race at the top and the first race last.

Data List 18
Juvenile Race z-Scores for Man o' War and Secretariat

Man o' War			Secretariat		
Race	Date	z Value	Race	Date	z Value
Futurity	13Sep19	-1.211	Garden St.	18Nov72	-1.484
Hopeful	30Aug19	-1.239	Laurel Fut.	28Oct72	-1.294
Union	23Aug19	-1.032	Champagne	14Oct72	-1.207
Sanford	13Aug19	-0.741	Futurity	16Sep72	-1.029
US Hotel	2Aug19	-1.310	Hopeful	26Aug72	-1.422
Tremont	5Jul19	-0.620	Sanford	16Aug72	-1.085
Hudson	23Jun19	-0.766	Alw	31Jul72	-0.987
Youthful	21Jun19	-0.839	Md Sp Wt	15Jul72	-1.327
Keene	9Jun19	-1.128	Md Sp Wt	4Jul72	0.857
Md Sp Wt	6Jun19	-1.325			
Mean:		-1.021			-0.997
F:		0.0028			
		$\alpha = 0.4634$			

The means, F score and t-test values for these two sets of z-scores are also given. The two shaded list entries represent likely inaccurate values. Man o' War's dubious z-score, -0.741, was for the Sanford Stakes in which he may have been severely misaligned at the barrier and suffered a late start. It was also his only career loss.

Secretariat's corresponding doubtful value, 0.857, is for his maiden race in which he was nearly knocked down shortly after exiting the gate. He lost that race, his only fourth-place finish, by 1-1/2 lengths after a spectacular recovery effort.

It was felt necessary to present the complete scores regardless of whether or not outliers might exist.

Although the variances of the two sets are significantly different, as shown by the F value of 0.0056, the one-tailed t-test, compensated for *unequal variance*, indicates that the two sets of scores do not differ significantly when taken as complete entities. The t-Test value of $\alpha = 0.4577$ is above 0.05.

When the NORMDIST function is run on both means, it indicates Man o' War's average z-score is in the lowest 15.36 percent of his peers, whereas Secretariat's mean z-score places him in the lowest 15.94 percent of his peers.

From these data, Man o' War and Secretariat should likely be judged equal as juvenile runners with respect to their peers.

As mentioned in BG, it only remains to determine whether Man o' War's peers were equivalent to Secretariat's. Common sense reasoning would say they were not due to foal crop size, for one. However, this assumption can be provisionally tested as follows:

Taking 2.5 percent of both foal crops gives 42 and 609, respectively, for the number of prospective single stakes winners, and, therefore, presumed better runners, in Man o' War's and Secretariat's crops.

Thus, for any given race, Secretariat's ratio for potentially facing top-level competition within his crop was, theoretically, 608/41 or 14.83 times more likely than for Man o' War.

However, this *does not prove* that Man o' War's competition was of lesser quality than Secretariat's.

Man o' War's average win margin, 2.30 lengths, combined with his general juvenile comment lines that he either 'won easing up,' won 'easily,' or was 'never extended,' give strong anecdotal evidence that he was either extremely talented, that he had no real challenges—or that both conclusions apply.

Secretariat's average juvenile win margin, 3.28 lengths, was greater than Man o' War's. However, Secretariat also bore less average impost, 119.33 lbs. per race versus Man o' War's 125.70 lbs., and none of his comment lines indicate that he was 'eased up' near the finish. Only one such comment out of nine, in fact, notes that he won 'easily.' However, his jockey, Ron Turcotte, repeatedly said that he never pushed him, and Turcotte rode him in 18 of his 21 career races.

So perhaps he had to try harder against his competition, despite a possibly more favorable combination of win margin and impost.

Peer Quality Factor

However, as an attempted response to the question of peer quality, an informal 'Peer Study' check was done on the horses finishing behind Man o' War and Secretariat for their respective juvenile races.

The *cumulative number of lengths* that the other finishers were behind the respective colts at the finish was used as a test criterion for the competition level each faced.

For ten juvenile races, Man o' War faced 60 total competitors of which 18 finished close enough to merit recorded lengths being published. For nine juvenile races, Secretariat faced 66 total competitors of which 35 finished high enough to merit recorded lengths.

Chapter Eight

A t-test of the two sets of combined lengths for all races for each colt tested *insignificant* with a t-value of α = 0.12. As usual, α = 0.05 or below is needed for significance.

Therefore, assuming *cumulative lengths behind* is equivalent to measuring difference in *competitor or peer quality* is *not supported* by this particular t-test result. We are back to square one on the subject.

The question of peer quality is elusive, and neither colt can apparently claim stiffer competition for their juvenile year than the other, based on this limited inquiry.

The tentative conclusion to the z-score study, therefore, is that Man o' War and Secretariat were, for all purposes, equal in running ability as juveniles, at least at six furlongs.

The following chapter gives the results of the same comparisons for the sophomore years of Man o' War and Secretariat as were presented in Chapters 7 and 8. Since all discussions of the sophomore results are based upon considerations nearly identical to the juvenile year, a more condensed presentation is given.

z-Scores and Time Adjustments Relative to the World Record

Chapter 7 included the results of comparing the six-furlong races of Man o' War and Secretariat to the existing world record of 2011 for that distance. The detailed explanation for how those results were derived is now presented.

The 2012 ARM (911) lists the world record holder for six furlongs as Twin Sparks. On November 21, 2009 he set the six-furlong record of 66.49 s at Turf Paradise as a six-year-old carrying 114 lbs.

The ARM also includes the 47 other fastest run times for that same distance, compiled for the years 1991 through 2011. Thus, an adequate sample is given by which to judge other horses.

The average and standard deviation for the fastest 48 six-furlong times calculate to: 67.29 s and 0.23 s, respectively.

We now wish to use z-scores for Man o' War and Secretariat to compare them on the *same scale* as the six-furlong world-record sample. The procedure is as follows:

The z-scores calculated previously for both colts for only their respective six-furlong races are used, plus the standard deviations specific to those same races.

Man o' War's average time for six furlongs was 72.21 s and his standard deviation was 0.72 s. Secretariat's equivalent values were 70.47 s and 0.42 s, respectively.

The z-score method allows a direct comparison of where Man o' War's and Secretariat's actual z-scores at six furlongs would place them *within the world-record sample*.

From Data List 18 it is found that the six six-furlong z-scores for Man o' War average -1.026. The three six-furlong z-scores for Secretariat average -1.133.

Since the world-record six-furlong sample has standard deviation 0.23, multiply 0.23 by the respective average z-scores of the two colts to determine where they would equivalently fall within the world-record sample.

For Man o' War, $(-1.026) \cdot (0.23) = -0.236$. Therefore Man o' War's equivalent time within the world-record sample is: [67.29 (the sample average) − 0.236] = 67.05.

Taking Man o' War's average z-score for six races gives an unbiased estimate of where he would place within the world-

Chapter Eight

record sample had he run on an equivalent track and under the same general racing tactics used by the 48 fastest horses and their jockeys, especially since we are operating under the postulate that ***overall racing ability within successive Thoroughbred generations is basically constant.***

Likewise, Secretariat's placement within the same world-record sample is at the point determined by his average six-furlong z-score multiplied by the world-record standard deviation. Its value is (-1.133) · (0.23) = - 0.261.

Thus, his estimated time referenced to the six-furlong world record is: 67.29 – 0.261 = 67.03.

These results are essentially identical. The only real difference is between the actual six-furlong standard deviations of the two colts. Those values are 0.72 s and 0.42 s, respectively. The standard deviations ultimately drive the ensuing simulations.

Therefore, 10 sets of 100 random numbers each were run using the above calculated values. The result was presented in Data List 17 of this chapter under the list heading 'z-Score/WR_a'.

The listed values total 504. This means that Man o' War is predicted to have 504 faster-time races of 1000—a near-perfect split with Secretariat—using the values just discussed for the averages and standard deviations at six furlongs of each colt.

It seems that a conclusion so provocative, considering that statistics gives only probable values for random fluctuations, indicates a good place to exit a chapter.

Indeed between 105 and 115 pounds the amount of weight carried by the horse is demonstrably unimportant. Beyond 115 pounds, the effect of added weight on performance is an individual thing.
~ Steven Davidowitz

Chapter Nine

Man o' War and Secretariat—Sophomore Year Comparisons

This chapter presents the same type data, in the same basic sequence, as Chapters 6 thru 8 did for the juvenile racing seasons of Man o' War and Secretariat.

Since much of the text in those chapters was explanatory and essentially the same explanations apply to the sophomore data, this chapter is more summary-oriented.

Basic Sophomore Year Data

During their three-year-old, or sophomore, racing seasons Man o' War ran eleven races to Secretariat's twelve.

Man o' War raced from May 18, 1920 to October 12, 1920. Secretariat raced from March 17, 1973 to October 28, 1973.

The major difference in the general makeup of these races was that Secretariat ran the final two races of his career on turf, whereas Man o' War never ran on turf.

Interestingly, Secretariat's penultimate career race was in the Man o' War Stakes at Belmont Park. It was a 12-furlong turf race, and his time was 2:24.80. Although that time set a new Belmont turf course record with a Speed Rating of 103, it was 0.80 s slower than his world record for 12 furlongs on dirt in the Belmont Stakes earlier that season.

Secretariat ended his career at Woodbine in Toronto, Ontario, Canada with a 13-furlong turf race. (DRF 279) It was the Canadian International, and jockey Ron Turcotte was not aboard Secretariat due to suspension for careless riding. Eddie Maple thus rode Secretariat to his last victory—by nearly seven lengths even as the burnished chestnut merely coasted under the wire following a 12-length lead at the head of the stretch.

Data List 19 presents side-by-side comparisons of the race results. Distances are given in both miles and furlongs; times are in seconds. The list entries are in chronological order from top to bottom—first race of each season at the top and final race at the bottom.

Data List 19

Man o'War		Secreteriat	
Distance	Time	Distance	Time
1.125, 9f	111.60	0.875, 7f	83.20
1.00, 8f	95.80	1.00, 8f	93.40
1.375, 11f	134.20	1.125, 9f	110.39
1.00, 8f	101.60	1.25, 10f	119.40
1.125, 9f	109.20	1.1875, 9.5f	113.00†
1.1875, 9.5f	116.60	1.50, 12f	144.00
1.25, 10f	121.80	1.125, 9f	107.00
1.625, 13f	160.80	1.125, 9f	109.35
1.500, 12f	148.80	1.125, 9f	105.40
1.0625, 8.5f	104.80	1.50, 12f	146.46
1.25, 10f	123.00	1.50, 12f, T*	144.80
		1.625, 13f, T	161.80

*T => Turf

†Secretariat's official Preakness time, as of June 19, 2012

Chapter Nine

As it was for their juvenile years, it is true for their sophomore years that the two colts ran just one common distance multiple times. That distance, 1.125 miles or nine furlongs, is strictly the only one that is directly comparable without undue bias, just as the six-furlong distance was for their juvenile season.

Correlations with Time for Nine Furlongs: Man o' War and Secretariat

The ten parameters in Data List 20 are from the DRF past performance sheet for Man o' War's nine-furlong races. Because he ran just two such races, the only possible correlations are 1.00 and -1.00. In four cases, such as for impost correlated with final times, division by zero occurs in the correlation formula and a 'div0!' warning is issued. This is one of the disadvantages of small sample sizes that cannot always be avoided.

Data List 20
Correlations with Time for Nine Furlongs: Man o' War

Parameters	Correlation	Coefficient of Determination
Call Point 1:	1.00	1.00
Post Position:	1.00	1.00
Track Condition:	Div0	no value
Overall Gain:	-1.00	1.00
Finish Margin:	Div0	no value
Field Size:	1.00	1.00
Impost:	Div0	no value
Speed Rating:	-1.00	1.00
Track Variant:	1.00	1.00
Rest Periods:	no value	no value

Unfortunately, the values in the list are unenlightening. Division by zero occurs whenever either the x or y variables being correlated all have the same value. In that case their corresponding standard deviations are zero.

To show this, enter some small number four or five times into a hand calculator in statistical mode (or into Excel). As you might expect, the mean of the numbers will equal their common value, the number repeatedly entered, and the standard deviation will be zero, by definition, since all numbers entered were equal and no variation exists.

Since the Pearson Product-Moment Correlation Coefficient formula contains the product of the x and y standard deviations in its denominator, that automatically makes division mathematically undefined, and the result is the ungainly looking '#Div0!'– presented in full ugliness here, but simplified for space savings in Data List 20.

Since only two nine-furlong time values are being correlated, either the second variable must automatically increase as the first increases or vice versa. In either case, 1.00 is the numerical result's absolute-value. The prefixed '+' or '-' simply shows the direction of the second variable relative to the first: same '+' or opposite '-'.

Data List 21, presenting Secretariat's equivalent parameter correlations, follows.

Chapter Nine

Data List 21

Correlations with Time for Nine Furlongs: Secretariat

Parameters	Correlation	Coefficient of Determination
Speed Rating:	-0.9824	0.9651
Call Point 1:	0.8889	0.7901
Overall Gain:	-0.7656	0.5861
Finish Margin:	-0.7633	0.5826
Track Variant:	0.7356	0.5412
Rest Period:	-0.6171	0.3808
Post Position:	-0.3647	0.1330
Field Size:	0.2289	0.0524
Impost:	-0.1759	0.0309
Track Condition:	see text	--------

The results of the sophomore-year parameter listings suggest that Man o' War's 3yo season does not strongly correlate with his 2yo season—due mainly to its small sample size. Since his correlations for the sophomore year are, by mathematical necessity, either -1.00 or +1.00, they cannot meaningfully be related to his juvenile year *or to Secretariat's* sophomore year.

Secretariat's sophomore parameters, however, can be related to his juvenile year since his nine-furlong sample size, N = 4, allows meaningful correlations. His correlations with time are given in Data List 21 where they are arranged in descending order by absolute value.

The five highlighted values are the most informative because they predict the highest percent of change in his times for nine furlongs. These percentages range from 96 for Speed Rating to 54 for Track Variant.

The same ten parameters for his juvenile season can readily be correlated with those in Data List 21. Each parameter is coded 1 through 10 and the ensuing positional change in parameters between racing seasons yields a meaningful correlation between the resulting two sets of coded values.

The rank correlation between the two parameter sets for these seasons is 0.2763. When this value is squared to find the common percentage of change across the two years, the resultant COD is 0.0763. This indicates that approximately 2 percent of the juvenile ranking for six furlongs is reflected in the sophomore ranking for nine furlongs. This is a weak relationship and indicates that, as Secretariat matured from his first to his second racing season, the relative importance of each parameter to his finishing times changed.

The two most significant position changes for these parameters were that Overall Gain gained five places (8th to 3rd) and Impost dropped six places (3rd to 9th).

Track Condition could not be rated for either year due to its constancy which resulted in the '#Div0!' indication already noted.

From its COD value, (0.0309), Impost was the least important parameter for Secretariat's nine-furlong races, whereas Overall Gain was for his six-furlong races, having COD = 0.0023.

Speed Rating was consistent for his two racing years. One would expect that numerous speed ratings approaching 100 would automatically assure fast finishing times and probable wins. This was true for Secretariat. His average speed rating for

Chapter Nine

four nine-furlong races was 93.75. He won two of these, placed in one and showed in the other.

He won all three of his six-furlong races with an average speed rating of 92.67—slightly lower than that for his comparable nine-furlong races.

Secretariat had a higher overall win percentage his first year versus his second year regarding total races (78% versus 75%). These percentages reflect his disqualification and resultant second place for his juvenile Champagne Stakes.

His trend in Speed Rating does not obviously contradict these percentages considering typical Thoroughbred data fluctuations.

Recall that Secretariat was ostensibly completely healthy for all his juvenile races. It has been documented that he was not so for the three races he lost during his second year, beginning with the Wood Memorial.

As was done for the juvenile year, t-test α values between the means for only the nine-furlong races are compared with the means of all other sophomore year races for each colt. Data List 22 presents the values. List items 1, 3 and 5 have units of lengths.

Data List 22
Test of Mean Values for 9 f versus All Races

Man o' War			Secretariat		
Parameter	9 f	All Races	Parameter	9 f	All Races
Call Point 1	46.80	38.90	Call Point 1	47.70	41.77
Field Size	5.50	2.55	Field Size	6.00	7.50
Overall Gain	0.94	13.28	Overall Gain	3.19	8.03
Impost	126.00	126.78	Impost	123.75	123.37
Finish Margin	1.50	17.67	Finish Margin	1.88	6.31
Post Position	4.00	1.89	Post Position	5.00	5.13

Data List 22 continued

Rest Period	18.00	14.33		Rest Period	20.25	16.14
Speed Rating	99.00	109.89		Speed Rating	95.00	98.00
Track Variant	9.00	7.89		Track Variant	11.50	9.13
Track Condition	1.00	1.22		Track Condition	1.00	1.88

As was true of the juvenile year, the t-tests for each colt suggest that no significant difference exists between the two sets of means. This implies that no special conditions were evidently active for the nine-furlong races with respect to the other sophomore races of either colt.

The final α level for Man o' War was 0.46 rounded to two decimal places. Secretariat's corresponding α level was 0.50. These two extremely close values nearly match the results for the juvenile years, 0.49 for both colts.

Thus, despite distinctly different running styles and parameter shifts, each colt apparently responded similarly to these ten defining condition parameters under which each ran during their career racing seasons.

This is a potentially strong argument to support a thesis that they possessed nearly identical ability.

Test for Significant Difference at Nine Furlongs

As was done for their juvenile times, a t-test was performed between the raw nine-furlong times for Man o' War and Secretariat. A significant difference was expected due to the arguably slower tracks run by Man o' War. Data List 23, representing all nine-furlong races for both colts, follows.

Chapter Nine

Data List 23

Man o' War				Secretariat			
Date	Track	Time	SR-TV	Date	Track	Time	SR-TV
5/18/20	Pim	111.60	97-10	4/21/73	Aqu	110.39	83-17
7/10/20	Aqu	109.20	101-08	6/30/73	AP	107.00	99-17
				8/04/73	Sar	109.35	94-15
				9/15/73	Bel	105.40	104-07

The sample size ($N = 2$) of Man o' War's races is just large enough to allow a t-test. Secretariat's sample size ($N = 4$) is also small, and caution must be applied in interpreting the result.

When the F test is run between the two samples, its value, 0.9864, indicates that the sample variances are uniform enough to apply the 'uniformity-of-variance' option for the t-test. A one-tailed t-test using that version gives $\alpha = 0.1347$.

This is somewhat surprising, especially given the small sample size. It indicates that even *the raw-score running times do not differ significantly*. Recall that a 0.05 t-test α level, or lower, is normally required for significance between independent samples—samples having no suspected pre-existing connections.

Technically this means that *no time adjustments* to Man o' War's nine-furlong data are needed to make them compatible with Secretariat's corresponding data.

Although adjustments will be made and their effects noted, this singular fact belies a real era effect.

Such postulated effects are never defined. They give the impression of being premature concerns of those who simply do not want comparisons between Man o' War and any other horse. Their protests are, it seems, based on a feeling that Man o' War will emerge the lesser in any comparison.

In fact, he does not, by any means, as the following results will demonstrate. These data *do not remotely indicate* his inferiority, and such fears are totally unfounded.

Ten simulations of 100 trials each from Excel's RNG using the means and standard deviations of the *unadjusted* nine-furlong raw data give Man o' War the following sequence of faster times: 18, 19, 19, 17, 23, 22, 26, 19, 25 and 23. Their total is 211. Secretariat, naturally, had the remaining wins from each set of 100 values.

This result stresses the fact that, even though a t-test indicates no significant difference, Man o' War still distinctly lags Secretariat in faster times.'

Although it appears more subjective than scientific to 'feel' that Man o' War would likely show a higher proportion of faster times than these results indicate, their face value is temporarily accepted just as are the conditions under which they were derived.

Now, however, another technique, *trapezoidal estimation*, is applied to determine by how much the unadjusted data must be changed to bring Man o' War into closer comparison with Secretariat.

Before proceeding with the trapezoidal adjustment method, the question of overall data normalcy is addressed since the results of t-tests and RNG simulations are generally based on assumptions of normalcy.

Shapiro-Wilk Normalcy Results: Ten Major Race Parameters

The ten chosen racing parameters were also checked for normalcy using the Shapiro-Wilk test on the sophomore date. Of the twenty separate values representing both colts, eight tested normal and the remainder tested non-normal.

In the 12 non-normal cases, it is likely that the combination of small sample size plus the numerous *repeat values* in the samples resulted in the non-normal Shapiro-Wilk probability.

The Shapiro-Wilk test invariably gives non-normal results when repeat values occur. As an example for both colts, Man o' War's Post Position parameter represented a sample size of 11. Of those 11 values, five 2s and four 1s occurred, just by the nature of the coding. That pattern more closely resembles a bimodal rather than a normal distribution.

Consequently, the Shapiro-Wilk test assigned a p-value, the probability that it was a sample from a normal distribution, of 0.0002—implying two chances in ten thousand.

Similarly, Secretariat's Impost sample contained 12 values, seven of which were 126 lbs. Its Shapiro-Wilk test gave the p-value 0.0026—implying 26 chances in ten thousand of it being from a normal distribution.

Non-normalcy of some variables is really non-problematic. In this case, the test was run more to highlight commonalities and possible patterns between the colts.

The *truly important variable*, the times for nine furlongs, tested normal for Secretariat. Its Shapiro-Wilk probability was 0.6934. That is above $\alpha = 0.05$ and is, in fact, strongly indicative of normalcy given the small sample size $N = 4$.

Man o' War had only two times for nine furlongs. Therefore, they could not be tested for normalcy since the Shapiro-Wilk requires a minimum of three values for its test.

It is virtually certain, however—given the test results of his other parameters, his times for six furlongs as a juvenile and the tendency of the Shapiro-Wilk test to readily assign non-normalcy to repeat values for small samples—that Man o' War's times for nine furlongs would test normally distributed for a larger sample.

Adjusting Running Times for Data Comparability

Note that no impost time adjustments are given to Man o' War for his nine-furlong races, contrary to what was done for his six-furlong races. The reason is readily seen by consulting the charts in Appendix J.

The charts contained therein for seven, eight, nine and 10 furlongs show negative correlation of time with impost increases. This implies that, as impost increases, over the typical range covered by the 50 or so fastest times for those distances—typically from 110 to 125 lbs.—running time decreases slightly.

This counter-intuitive, if not contradictory, observation finds support from experienced handicappers such as Davidowitz, whose quote was used as the heading for this chapter (122). Impost often correlates inconsistently with times.

The impost issue now at least marginally resolved, the same general time-adjustment methods are applied to Man o' War's and Secretariat's times for nine furlongs as was done for the six-furlong times.

Therefore, in keeping with the initial adjustments for their juvenile year, the *standard of track speed* to which both Man o' War's and Secretariat's times for nine furlongs were adjusted is the current North American Dirt Track Record as published in the American Racing Manual, 2012 edition. (911)

The ARM 2012 gives that record, as of 2011, as being held by Simply Majestic, a four-year-old carrying 114 lbs. It was set

Chapter Nine

at Golden Gate Fields in Berkeley, California on April 2, 1988. Simply Majestic's time was 1:45.00, or 105.00 s in the format preferred for mathematical calculations.

Like the majority of colts and fillies listed in the ARM as holding world records or near world records, Simply Majestic is not among the elite horses included in the DRF 2005 Champions publication, even with the very respectable career record of 18 wins, 4 seconds and 7 thirds out of 44 career starts.

This fact argues strongly for the premise stated earlier regarding uniformity of Thoroughbred ability across successive generations.

That is, sustained speed is genetically determined and delimited within the species regarding the highest levels it may attain, given additional proper encouragement and training. Only a small percentage of foals within any given crop are, *a priori*, capable of attaining it.

Furthermore, if a horse setting a world-level record never goes beyond that record to achieve acclaim, it must have been running at least near the species apex, or minus- three-sigma limit, when the record was set.

In other words, simply because a given horse *can* attain and maintain a certain pace in a given race does not necessarily mean that he will do it more than once—witness Dark Star's win by a head over the great Native Dancer in the 1953 Kentucky Derby—he may or he may not. The truly great generally duplicate the achievement multiple times.

If a horse establishes a world record for *any* given distance, and especially the longer the record stands, it is arguable that the record is somewhere near the aforementioned species apex. It is, if you will, a kind of Thoroughbred 'absolute zero' by which subsequent times may be compared.

Therefore, since world records may be viewed, even though set 92 years after Man o' War ran and 39 years after Secretariat's triumphs, as the lowest potential time for a given distance attainable by an equine, it is proper to compare Man o' War's and Secretariat's times by reference to such a record.

The Sequence of Time Adjustments

The following factors were used, as was done for six furlongs, as guides to adjusting Man o' War's and Secretariat's run times for nine furlongs: a) the average Speed Rating, or Track Variant, for the track on the day the race was held, b) the actual Speed Rating of each colt being compared and c) the existing track records *where and when* each colt ran.

One example of the adjustment method for the nine-furlong races for Man o' War is presented. The method should them be clear with no further examples.

Recall that impost adjustments *are not given* for the nine-furlong races because inconsistent and low correlations exist between impost and world record times, as detailed in Appendix J.

Man o' War ran two nine-furlong races as a sophomore. His times for these races were, using decimal format, 111.60 s for the Preakness on May 18, 1920 and 109.20 s for the Dwyer Stakes on July 10, 1920. His average time and standard deviation for these two races combined were: 110.40 s and 1.70 s, respectively.

Note that the Dwyer Stakes was *probably the only sophomore race* in which Man o' War was truly challenged, that being by the colt John P. Grier. The race's comment line states "hard

Chapter Nine

ridden, drew away." Man o' War eventually won the race by 1-1/2 lengths.

Secretariat ran four nine-furlong races as a sophomore. His times, again using decimal format, were: 110.39 s, 107.00 s, 109.35 s and 105.40 s. His raw (unadjusted and rounded) time average and standard deviation for these races are: 108.04 s and 2.26 s, respectively.

He lost the first and third of these, the Wood Memorial and the Whitney, and won the second and fourth, the Invitational and the Marlboro Cup Invitational. The times given for Secretariat's losing races were adjusted, by the method previously explained, to reflect the margins behind by which he finished.

His comment lines for the two losses state: "wide, hung' for the Wood Memorial and "weakened" for the Whitney. His less than robust physical state for these races was noted previously.

His comment lines for the two wins do not indicate that he was obviously "asked" to run. Indeed, his jockey, Ron Turcotte, repeatedly stressed never asking him, generally saying that he wanted to save him for the next race. (ESPN)

From Secretariat's average and standard deviation for the nine-furlong races, it is easily determined that Man o' War would not need four or five seconds attributed to slower tracks in order for his average time to match Secretariat's. Their nine-furlong averages only differ by 2.36 s, as noted in the Prologue.

Thus, if his advocates wish to continue arguing for non-comparison due to era effects, then the presumed effect for his sophomore year was no more than 2.36 s, at least for the nine-furlong races.

Raw Data Comparison
Simulations using Raw Data

When a second, corroborative set, of 100 simulations were run 10 times using only the *raw data* (original run times) for both colts, Man o' War was credited the following sequence of 'wins' by the RNG: 25, 15, 20, 19, 21, 19, 18, 21, 24 and 26. The total is 208 projected fastest times out of 1000 simulations. This nicely matches the 211 noted previously for the first such set.

It cannot be overemphasized that *known random drawings* from a *known normal distribution* will result in a *distinct range* of expected outcomes and not simply one basic result, as shown for the preceding sequence of 211 faster times. In this case the lower simulation limit is 15 and the upper limit is 26. The range, 11, is the difference between these two values.

The same general fluctuations are found in any simulated results applied to time adjustments. In this sense, there is no such thing as a precise prediction using statistical methods.

t-Test Results and a Reference Baseline

A basis for comparison can be established by which to judge the effectiveness of the remaining time adjustments. The most direct way to determine this baseline is by using the t-test between the original, unaltered, nine-furlong times for both colts.

The result is that the difference in variances tests non-significant with a two-place F ratio of 0.99. The one-tailed t-test, supported by the non-significant F test result, gives a non-significant difference between the means with $\alpha = 0.13$.

That is, if the null hypothesis states that *no difference* between the means exists, the t-test *confirms this assumption*

Chapter Nine

by indicating that 13 chances in 100 (0.13) exist of *randomly drawing* two samples having the given spread in means (2.36 s), either from the same population or from two populations having equal means. This is considered a high enough probability statistically to judge that the samples do not differ significantly.

Adjustment Method 1: Comparison of Track Records

Man o' War's and Secretariat's original nine-furlong times, Speed Ratings, Track Variants and existing track records are listed below. The abbreviations should be obvious, being simply the first letters of the corresponding words. $Time_0$ denotes original times.

All numerical values in Data List 24 have units of seconds.

Data List 24

Man o' War			Secretariat		
$Time_0$	SR-TV	TR	$Time_0$	SR-TV	TR
111.60	97-10	111.00	110.39	83-17	106.40
109.20	101-08	109.40	107.00	99-17	106.80
			109.35	94-15	108.00
			105.40	104-07	106.20

To adjust Man o' War's times relative to Secretariat's the assumption is made that the existing track record for both colts is the best single indicator of true track-speed differences that may have affected their performances.

In many cases, however, a track record is set by just a good horse running a superb race—the "race of his life"— as has now been repeatedly emphasized.

The record time, however, at least indirectly represents the effect of track speed on even the best performance seen at the

203

track. Therefore, it is the *best logical value* to use for comparisons across large time or generation gaps such as between Man o' War and Secretariat.

Since Secretariat ran four nine-furlong races to Man o' War's two, a direct comparison, race by race, cannot be made. Therefore, the most stable statistic, the *average* or mean value, will be used to adjust Man o' War's times to Secretariat's, as it was for their six-furlong races.

When the average of the four track records where Secretariat ran is calculated, the result is 106.85. That value is then subtracted from *each* of the existing track record times where Man o' War ran.

A single example is now given of the first time adjustment for Man o' War. Since his time for the first nine-furlong race was 111.60 s and his Speed Rating for that race was 97, this means that his time was 100 − 97 or 3/5 s (0.60 s) slower than the standing track record. Therefore, subtracting 0.60 s from 111.60 s gives 111.00 s for the track record (TR) where Man o' War ran.

Man o' War's nine-furlong track variants were 10 and 8. One-half the time value of each is, therefore, given to him. This translates to 1.00 s for the first race (5/5) and 0.80 s (4/5) for the second race. This same technique was used with the juvenile data.

The *average* track record value for Secretariat's four nine-furlong races was 106.85 s. There is good probability that the difference between the two track records closely indicates the actual difference in track speeds he and Man o' War experienced.

Therefore, if the best time a horse had run up to that time on Man o' War's track (Pimlico) was 111.00 s for nine furlongs and the best *average time* other horses had run on Secretariat's tracks, even though it was 39 years later, was 106.85 s, then direct subtraction indicates that Pimlico in 1920 was possibly 4.15 s slower than the average track Secretariat ran.

This is a reasonable assumption because a separate linear regression performed on Preakness running times versus foal crop size (independent variable) from 1925 through 1973 generated the equation -0.00023 x + 114.12.

When the values of Man o' War's and Secretariat's foal crops from 1922 (2,352) and 1970 (24,361) are substituted for 'x' in this equation and the calculations are completed, the resulting estimated values for average running times are 113.58 s for 1925 and 108.52 s for 1973. The difference between these predicted values is 5.06 s.

This provides a decent check against the 4.15 s figure because the equation's COD is 0.5485. This COD value means that the equation predicts changes in running time with about 55 percent accuracy. This, in turn, implies that the actual times, given the complementary chance of 45 percent prediction error, are between approximately 3.48 s and 9.18 s (from the equation $\hat{y} \pm 0.45\hat{y} = 5.06$ s).

Since the 9.18 s value violates the "common sense" criterion, a reasonable expected value is somewhere between 3.48 s and 6.33 s, the latter value being the midpoint of the entire error range.

Therefore, 4.15 s being within that range, it is reasonable that the Pimlico track was about four seconds slower when Man o' War ran versus when Secretariat ran. It is also *possible*, based on this brief analysis and the nature of data fluctuations, that the

Pimlico track *was about* 4.15 s slower than typical tracks where Secretariat ran.

When the equivalent calculations are made for Man o' War's second nine-furlong race, the difference between his theoretically inherent track speed (Aqueduct) and Secretariat's average inherent track speed calculates to 2.55 s.

This naturally implies that Aqueduct was about 1.60 s faster than Pimlico. The average of the two differences is 3.35 s.

Therefore, 3.35 s is the time adjustment value that is given Man o' War for each of his nine-furlong races. Note that the *same result is obtained* in terms of the final *average* nine-furlong time attributed to Man o' War if 4.15 s and 2.55 s were *first subtracted* from his respective raw times and *then* the average of that result was taken.

Consider also that 3.35 s is certainly compatible with the statement made by Nack (401) in his descriptions of Secretariat's time for the Belmont. Secretariat's recorded time at the eleven-furlong mark in that race was 3 seconds faster than Man o' War's when Man o' War established the Belmont record in 1920 – 131.20 s vice 134.20 s.

In 1920 the Belmont was eleven furlongs vice twelve furlongs beginning in 1926. Therefore, on this basis, it is reasonable to assume that a 3.35 s time credit to Man o' War is equitable.

It is appropriate to provide a baseline perspective, as did Chapter 7, Figure 7, for the range of logical time values which can be attributed to Man o' War without showing undue bias to Secretariat. Figure 9 presents six such values.

Chapter Nine

Figure 9

Logical boundaries for Man o' War's 9-f Time Adjustments

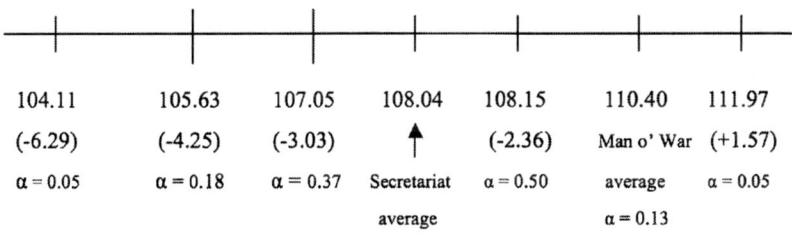

Reading from right to left along the line, the rationale for each point is as follows, where the units are seconds for each parenthetical number.

If 1.57 s is *added* to Man o' War's original nine-furlong times, it places them at the upper boundary level, $\alpha = 0.05$, where they significantly differ from Secretariat's. At and beyond that level, Man o' War has *decreasing chances* of ever having a faster simulation time than Secretariat.

The second point leftward, 110.40, represents Man o' War's original average time for two nine-furlong races. At this level, for a typical simulation, he receives an average of 19 of 100 faster times than Secretariat. Their two sets of nine-furlong times do not differ significantly at this level, having $\alpha = 0.13$.

The next adjustment level, -2.36, indicates the difference between the average nine-furlong times for both colts. That is, subtracting Secretariat's average nine-furlong time, 108.04 s, from Man o' War's average time of 110.40 s leaves the remainder 2.36 s.

The next lower level, adjustment -3.03, represents the correction to Man o' War's times using the trapezoidal method (to be discussed) for nine furlongs. When 3.03 s is subtracted from each of Man o' War's times, alpha's value relative to Secretariat's unadjusted times is 0.37. They are, therefore, not significantly different.

At the next lower adjustment level, 4.25 s are deducted from each of Man o' War's times. This subtraction results from combining the Track Variant adjustment with the difference in times between the track records of both colts. Alpha's value, now lowered to 0.18, is still not significant.

The lowest adjustment level, resulting from lowering each of Man o' War's times by 6.29 s, represents the lower $\alpha = 0.05$ boundary. At or below this level, nearly all Man o' War's assigned simulation times were *faster than* Secretariat's. A summary of these time sequences follows.

Simulation Results of the Adjustments

When 100 RNG simulations are run for the compensations just discussed, the average number of faster times for Man o' War are: seven at the upper or rightmost $\alpha = 0.05$ level, 19 for his original times, 53 when 2.36 s is subtracted from each of his times, 59 when 3.03 is subtracted, 81 when 4.25 s is subtracted, and 92 at the lower $\alpha = 0.05$ level.

Thus, Figure 9 provides a convenient guide to the number of faster times in 100, or any number in fact, that can be expected for any given adjustment.

From Figure 9 it is easily determined that the lower $\alpha = 0.05$ level of significance is probably biased against Secretariat in light of the known facts.

Chapter Nine

Even staunch devotees of Man o' War would probably not insist that tracks were over six and one-quarter seconds slower when he ran vice when Secretariat ran. The ARM records, among others, would contradict such statements.

The next-to-last adjustment, -4.25 s, refers only to Man o' War, leaving Secretariat's times unadjusted. Secretariat's times must also be adjusted to at least compensate him for Track Variant effects. With unadjusted times, Man o' War gains 723 faster times of 1000 than does Secretariat.

As was the case for six furlongs, Secretariat's track record for nine furlongs cannot be used for adjusting his times because that would essentially cancel any gain Man o' War received relative to the same standard.

The only equitable alternative is to use Secretariat's Track Variants and credit him with *one-half* of the speed value they represent, as was done for Man o' War.

As indicated previously regarding the six-furlong time compensations, Appendix H gives supporting data for assuming a 50-50 percentage split of Track Variant into track speed effects and other effects such as jockey decision making and foal quality.

Secretariat's Track Variants for his nine-furlong races were: 17, 17, 15 and 7. Allotting 50 percent of each of these numbers to track speed effect gives: 8.5/5, 8.5/5, 7.5/5 and 3.5/5, respectively, with these fractions in terms of fifths of one second.

The corresponding decimal values for the fractions are: 1.7, 1.7, 1.5 and 0.7, respectively. These are the times in seconds by which Secretariat's run times *would normally be adjusted* since they represent the assumed portion of the Track Variant affecting speed.

Secretariat's final adjusted times are: 105.40 – 0.70 = 104.70; 109.35 – 1.5 = 107.85; 107.00 – 1.70 = 105.30 and 110.39 – 1.70 = 108.69. The mean and standard deviation for these four values are: 106.64 and 1.93, respectively.

When these values for both colts are used as RNG parameters in 1000 nine-furlong simulations, Man o' War's faster times in 10 successive trials of 100 each were: 62, 49, 53, 49, 61, 55, 56, 53, 60 and 55, with complementary faster times going to Secretariat. The sequence totals 553.

A t-test between the two resulting sets of adjusted run times has $\alpha = 0.40$, indicating that they are not significantly different, based on $\alpha = 0.05$ as the criterion.

It can be seen that exceeding the Figure 9 boundaries in either direction give t-test values for α lower than 0.05 and, therefore, that increasingly significant differences automatically follow.

This is not saying that it was impossible for Man o' War to have posted faster times than Secretariat. The real question is—if Man o' War's times were actually faster than Secretariat's, how much faster need they be *at minimum* so that an *unbiased estimate* reveals this value without unduly penalizing Secretariat?

The next section offers a possible solution to this vital query.

The Time Adjustment Giving Man o' War the Obvious Advantage

Most betting people would agree that winning a single match does not fairly judge the 'loser' for having an atypical performance. Thus, major sports such as professional baseball and basketball establish a "best-four-of-seven series" to determine the season's champion.

Chapter Nine

Four wins out of seven events essentially represents a win percentage of 57. One can tentatively use this number and say that if a time adjustment for Man o' War resulted in his being assigned a minimum of 57 or more out of 100 faster times via random number simulation, then *his record could be judged better* than Secretariat's.

Notice that the expression 'his record could be judged better . . .' and not 'Man o' War could be judged better . . .'was purposefully used.

This expression was applied to the result because, as stated in my other books on Thoroughbred data comparison, statistics is *a priori* incapable of judging the inherent *quality* of any creature. Statistics only compares inanimate data values—period, exclamation point and end of discussion!

This admonition having been firmly offered, the question still lingers concerning what *ultimate and reasonable* number of faster times out of 100 would *best define* the expression 'having a better record for a given distance'? The horses involved are put aside at this juncture.

To help answer this question, 100 Bernoulli random trials were generated to simulate tossing an unbiased coin. Arbitrarily, '0' denoted a head and '1' denoted a tail. The Excel function for Bernoulli trials (in which only two possible outcomes are generated) is accessed via the CRITBINOM function. This foreboding looking acronym stands for 'Critical Binomial.'

It was decided that *a given colt's record must show more 'favorable' outcomes than what random fluctuations alone would typically generate in a series of unbiased coin flips to qualify as truly showing superiority.*

The result of the Bernoulli trials was that a range of 'heads' from 49 to 59 was generated when 10 trials of 100 'flips' each were produced.

Therefore, 'better than chance' when winning events such as races is at stake should require *more* than the upper level of 59 in the Bernoulli trials.

Since just over 68 percent of the normal distribution is contained between ± one standard deviation on either side of the mean, and since that level reasonably exceeds 59, the final time-adjustment level that gave Man o' War 70 percent, or more, faster random times for a given distance was selected. This level is intended to err conservatively.

Therefore, 70 percent was defined as the 'win' level needed to claim that one record was consistently better than another.

The NORMINV function was used for this simulation. The boundary between the lower 30 percent and the upper 70 percent under Secretariat's normal distribution for his adjusted nine-furlong running times was used as reference. This distribution has a mean value of 106.64 and a standard deviation of 1.93.

The result was that Man o' War's nine-furlong times, when each was adjusted downward by 4.95 s, with Secretariat's adjustment remaining at -1.40 s, gave him 654 of 1000 faster simulation times at a rounded α level of 0.28.

His corresponding sequence of ten values for 100 simulations each was then: 68, 69, 73, 62, 59, 68, 75, 54, 66 and 60. The range of these values is: 75 − 54 = 21. This range is, as shown previously, to be expected when random simulations are used.

At this level, his adjusted nine-furlong times become 106.65 s and 104.25 s. His resultant average nine-furlong time is 105.45 s and his standard deviation remains 1.697. By comparison, the corresponding 2011 world-record nine-furlong time is 105.00 s.

This adjustment is deemed to be Man o' War's likely *minimum lower limit.*

Thus, Man o' War's adjusted average is just 0.45 s higher than the 2011 nine-furlong world record. That does not seem bad for a horse from a presumed different era!

If his times, using the 4.95 s adjustment, are compared with Secretariat's *unadjusted,* original times, then Man o' War's simulated faster times increase to an average of 82 of 100. This is nearly the reverse of the situation when Man o' War's unadjusted times are run against Secretariat's unadjusted times. And so a kind of symmetry, beloved of physicists, follows.

It is probable that 4.95 s represents a *greater time adjustment than merited* by the tracks Man o' War ran. The existing evidence indicates this is true.

As supporting argument, consider that the following numbers represent the actual average time differences between Man o' War's and Secretariat's decades for four quintessential American races: 4.35 s, 3.00 s, 4.82 s and 4.18 s.

Those times represent the changes, *without including foal crop effect,* for the Travers, Belmont, Preakness and Kentucky Derby, respectively—each compared for the five to ten years closest to 1920 (Man o' War's 'era') and the decade surrounding 1973 (Secretariat's 'era'). Their average is 4.09 s and their standard deviation is 0.77 s.

Minus three such standard deviations allows that as low as 4.09 s − 3·0.77 s = 1.78 s could be an equitable time subtraction for Man o' War, whereas as high an adjustment as 4.09 s + 3· 0.77 s, or 6.40 s time subtraction could also be equitable.

The higher adjustment level is 0.11 s lower than the lowest suggested level in Figure 9. Therefore, only a person with a biased sense of propriety would, in my opinion, suggest the latter adjustment.

It is, however, interesting that the two numbers are close, being arrived at by disparate routes. It helps establish the internal consistency of the arguments leading up to them.

In GHA, it was shown that foal crop could account for the following percentages of running time changes: Travers—87%; Belmont—74%; Preakness—97% and Kentucky Derby—79%. However, in the spirit of liberalism, assume that a 50-50 percent split is actually true, as previously used for time adjustments related to Track Variant.

The actual time effects these tracks have, under a 50-50 assumption, are: Travers (Saratoga)—2.18 s; Belmont (Belmont Park)—1.50 s; Preakness (Pimlico)—2.41 s and Kentucky Derby (Churchill Downs)—2.09 s. The overall average is 2.05 s.

This estimated, average cross-decade time difference for the above four races is reasonably close to Farley's estimate regarding the likely time difference between Man o' War's tracks (almost two seconds at eight furlongs) versus the tracks of more modern runners—referring to the year 1962 when his book was published. (337)

Therefore, since 2.05 s is obviously less than the 4.95 s adjustment just discussed, it and *any lesser time adjustment* automatically mean that Man o' War's number of simulated faster times over Secretariat *would not be significant*—no more so than the fluctuations expected in coin tossing as considered on a Bernoulli basis. This result begs a conclusion to which it is left for reader interest and emotional predilections to decide.

Chapter Nine

It is also consonant with the author's avowed intent not to use statistics for judging the inherent quality of one runner over another.

The same detailed analysis was not used for the six-furlong times because colts that young are even more difficult to compare equitably than are three-year-olds. They are in early developmental stages, often termed 'babies,' whereas three-year-olds are in middle development or somewhat beyond.

It seems appropriate to end this chapter with a reminder. That is, all Thoroughbred colts,—Man o' War and Secretariat particularly—deserve just comparisons, if they are to be compared at all.

If investigators do not follow this dictum, or even dogmatic assertion if you will, then I agree totally with the admonition of not comparing horses from different eras, an 'era' in that case simply being another term for different decades within the given century.

It is not, however, a true, undefined era difference that prevents comparison. Rather, it should be an internal attitude serving constantly as the statistician's moral and ethical compass and voice of reason that urges comparisons to proceed with respect.

But just comparisons are sometimes nearly impossible to determine—due as much to gaps and unanswered questions in the data records, rather than to the imagined influence of an *ad hoc*, itinerant era poltergeist.

Facts of this general kind may be the actual basis for Shakespeare's admonition used at the beginning of the Prologue regarding comparisons being odorous—or 'odious' as some prefer the quote, even though the original be thereby distorted.

His only point of reference is himself.
~ Charles Hatton

Chapter Ten

Adjustment Method 2: The Trapezoidal Time Adjustment

The same technique is now applied to the sophomore year data for nine furlongs as it was to the six-furlong data. The preceding chapter previewed the nine-furlong data.

This method requires using linear regression combined with the trapezoidal area beneath the regression line and included between the appropriate distances along the horizontal or x-axis, the axis of the independent variable, distance.

In Chapter 7 both Man o' War's and Secretariat's linear trend lines were adjusted downward until the areas beneath each equaled the area beneath the world-record regression line which extended from five furlongs to thirteen furlongs.

Since we now wish to compare the areas centered on nine furlongs, the baseline range is set between eight and one-half and nine and one-half furlongs, the distance representing the so-called 'height' of the trapezoid.

Figure L2 of Appendix L illustrates the method. The only difference between the six-furlong and nine-furlong examples lies in the numerical values of the *six endpoints* involved—two each for the world-record line, for Man o' War's linear regression and for Secretariat's linear regression.

The endpoints for the world-record linear regression come directly from the data supplied in the 2012 ARM for North

217

American Dirt records. (911) The respective times, for eight and one-half and nine and one-half furlongs—as used for this adjustment—are 96.40 s and 112.40 s.

The endpoint time values for Man o' War are: 104.37 s and 116.79 s, whereas Secretariat's corresponding values are 101.13 s and 113.97 s.

Times are in seconds, and each time comes directly from Excel's FORECAST application applied to each colt's corresponding linear regression.

Man o' War's 3yo regression equation is: 99.35 x -1.19 and Secretariat's is: 102.71 x − 8.00, based on distance (x) in miles and fractional miles. All equation numbers are rounded to two decimal places.

Secretariat's regression was derived using his official revised time of 1:53.00 or 113.00 s for the Preakness Stakes, now officially approved by the Maryland Racing Commission on June 19, 2012. (Associated Press 1)

Data List 25 repeats Data List 15 from Chapter 7. All numerical values represent the equivalent areas beneath the corresponding reference lines—the linear regressions cited above.

Technically, these areas have squared units of furlong-seconds associated with them, analogous to the use in physics of work being represented as an area under a force-distance graph, as explained previously. For that case, the resultant units of work are, in the MKS System, Newton-meters, representing force multiplied by distance.

It is obviously much simpler to drop the units because the numerical value is the important focus of these calculations, so long as all calculations remain consistent.

Chapter Ten

Data List 25

	Juvenile Year				Sophomore Year			
Man o' War	Secretariat	World	Delta		Man o' War	Secretariat	World	Delta
72.20		67.02	5.18		110.58		104.40	6.18
	70.49	67.02	3.47			107.55	104.40	3.15
		Final:	-1.71				Final:	-3.03

The data list shows a difference in magnitude existing between the areas for the juvenile and sophomore years. This was expected. It is partly due to the small sample sizes upon which both regressions were based and the fact that the regression CODs, although above 0.99 for each colt's sophomore year, nevertheless introduce error into any estimate.

The error is partly anticipated by the standard error of estimate (SEE), item #6 in the LINEST example chart of Appendix D. For this case, Man o' War's SEE is 1.66 s and Secretariat's is 1.90 s with decimal rounding to two places.

It was previously stated that this particular technique was considered exploratory in the sense that, if the results seemed unreasonable, little reliance would be placed on them. That problem did not arise.

For the juvenile case, subtracting the suggested 1.71 s from each of Man o' War's six-furlong times resulted in an average of 70.50 s and a standard deviation of 0.72 s.

If Farley's estimate (337) of about two seconds per mile actually represented the speed of Man o' War's tracks compared with Secretariat's, and if "about" is interpreted as meaning "within the standard of time measurement" existing in 1920 (one-fifth of one second), then it is logical to expect 1.8 s as the difference between the two track speeds.

In that case, the proportional amount for six furlongs is 1.35 s. The current value of 1.71 s seems realistic given the data fluctuations associated with Thoroughbred racing. It is, after all, less that 2/5 s, and this is not rocket science.

Thus, 1.71 s was the adjustment value derived for six furlongs using the Trapezoidal technique.

As mentioned previously, the more economical way of adjusting Man o' War's trapezoidal data is to find the difference between his and Secretariat's adjusted area values relative to the world-record area and *then* directly subtracting those two values. The result is 6.18 – 3.15 = 3.03.

Subtracting 3.03 s from each of Man o' War's nine-furlong times results in an adjusted average and standard deviation of 107.37 s and 1.70 s, respectively.

When 1000 random simulations using these values are run against Secretariat's unadjusted nine-furlong values, the result is that Man o' War is credited with 572 of 1000 faster times than is Secretariat. The derived sequence of 10 trials having 100 values each was: 62, 51, 57, 50, 63, 58, 59, 53, 60 and 59.

The data analysis portion of the book now concludes with a discussion of the sophomore z-scores of each colt and their resultant placement relative to the 2011 world record for nine furlongs.

The Relationship of z-Scores to the Nine-Furlong World Record

The same general method of assigning placement to both Man o' War and Secretariat within the world-record time distribution for nine furlongs is followed as it was for six furlongs.

Chapter Ten

The average of the fifty-two fastest nine-furlong times, published by the 2012 ARM, is 106.33 s with standard deviation 0.35 (935)

Data List 26 presents the complete z-scores for both Man o' War and Secretariat. The scores for nine furlongs are highlighted.

Data List 26
Complete sophomore z-Scores for Man o' War and Secretariat

Distance	Mow o' War	Distance	Secretariat
8	-0.701	7	-1.393
8	-0.713	8	-1.043
8.5	-0.936	9	0.755
9	-0.940	9	-1.005
9	-0.705	9	-0.320
9.5	-1.064	9	-1.045
10	-0.814	9.5	-1.063
10	-0.714	10	-1.074
11	-0.706	12	-1.409
12	-0.710	12	-0.566
13	-0.708	12	-1.216
		13	-1.466

In the list, all distances are in furlongs. Distances are not in their original chronological race order. They are sorted from low to high. All z-scores are rounded to three decimal places.

As for the six-furlong races, the average z-score for each colt at nine furlongs was first obtained. Man o' War average is -0.792 and Secretariat's is -0.904.

Each average was then multiplied by the world-record standard deviation, 0.35.

Man o' War's result was -0.277 and Secretariat's was -0.316. The intentionally omitted units are seconds.

These two values, or *indicators*, determine the placement of Man o' War's and Secretariat's averages relative to the mean of the world-record distribution.

Man o' War's indicator places him at 106.33 − 0.277 = 106.05 relative to the world-record mean, and Secretariat's places him at 106.33 − 0.316 = 106.01 relative to the world-record mean.

Simulations run with Excel's RNG need both an average and standard deviation entered before random times can be assigned. The actual nine-furlong standard deviations for both Man o' War and Secretariat were used for this application. Those values are 1.697 and 2.258, respectively.

The ensuing simulation indicates that Man o' War receives 494/1000 faster times. The breakdown, in groups of ten simulations of 100 values each, is, in the order generated: 50, 57, 35, 43, 49, 48, 52, 51, 56 and 53.

This typical random sequence, having minimum value 35, maximum value 57 and range 57 − 35 = 22, again illustrates the nature of random processes and emphasizes how theory indicates or intimates what practice would never allow to be shown.

Namely, in real life—even though perfect clones of Man o' War and Secretariat were produced in the foreseeable future, including particularly their extraordinary running abilities—there is no way in which they would ever be allowed to run *two* match races, much less five, ten, one hundred or a thousand—even if they were physically able to do so and enough time existed for the events to occur.

It would be a cold day in Calcutta before such an occasion arose.

Chapter Ten

Real horsemen simply do not jeopardize horses in that manner, nor do they wish their horse to come out the loser. And one horse would, invariably and by definition, be the loser of any given match. Photo finishes are extremely rare.

We see, from the results of every simulation discussed herein, that relatively wide fluctuations always occur within that technique, just as they would in real life with its myriad of interacting, random and even unknown variables affecting the outcome of each contest.

That, in fact, is a main reason why random simulations are invaluable adjuncts to comprehensive Thoroughbred data analysis and comparison—they intimate *by their variations* how difficult it is for a Thoroughbred to register an unbeaten career.

It will suffice here, perhaps unpalatable as it may first seem, to realize that Man o' War and Secretariat should be judged essentially equal from the z-scores and the previous analyses. The data virtually insist upon it.

Man o' War's "winning" 494 of 1000 simulated races really is no better—and no more cause for rejoicing by his devotees—than would be the occasion of flipping a coin 1000 times and having either 494 heads or 494 tails show. That number is simply well within the bounds of expected random sampling "fluctuation error."

In fact, if one closely inspects the breakdown results of this latest set of ten simulations of 100 values each, one realizes that Man o' War would have tied the first, barely won the eighth, lost the third through the sixth, and won the second, seventh, ninth and tenth.

Such results do not remotely meet the criterion established earlier—that, to appear definitely "better than" a second competitor, the expected fluctuations of a sequence of binomial trials should assign at least 70 percent of favorable outcomes to one of two subjects.

The preceding discussion begs the question of just how many seconds must be deducted from Man o' War's nine-furlong times so that he *does* gain roughly 70 percent or more of the total simulation faster times.

This answer is found via linear regression. If a LINEST is performed using the average number of random faster times out of 100 assigned *to each* of the separate time deductions shown in Figure 9 from Chapter 9, a linear regression with a COD of 0.996, thus having over 99-percent predictive accuracy, is obtained.

The dependent variable for this regression is *number of random wins* and the independent variable is the resulting amounts of *time deducted*. The values for the respective dependent/independent variable pairs (read in that order) for the regression are: (20/0, 49/2.36, 57/3.03, 72/4.25 and 92/6.29), where the first value of each pair represents the rounded number of average random "faster times" obtained for Man o' War, and the second represents the number of seconds subtracted from his average time.

The resultant linear regression equation is: $\hat{e} = 11.524\ \mathbf{t} + 21.263$. In this equation, $\hat{e}$ (e-hat) represents the expected average number of predicted random "faster times" and **t** represents the time subtracted from each of Man o' War's nine-furlong times.

Chapter Ten

By setting $\hat{e} = 70$ in this equation and solving for **t**, one finds, within rounding error, that **t** = 4.23 s typically gains Man o' War and average of 72 faster times (integer rounding) of 100 versus Secretariat's *unadjusted* time values.

The sequence of ten groups of 100 in 1000 total, based on this particular derivation, was: 74, 66, 75, 63, 75, 70, 74, 79, 73 and 74. The set's average is 72.3 and its standard deviation is 4.72. The range is 79 – 63 = 16.

The value 4.23 s is the *approximate* answer sought. It is approximate precisely because the sequence just presented for the 1000 trials fluctuates randomly.

Recall that the previous chapter also presented a 70-precent win level for Man o' War assuming that 4.95 s was subtracted from each of his nine-furlong times *and* 1.40 s was subtracted from each of Secretariat's.

The essential difference between the two comparisons is the 1.40 s credit to Secretariat in the former simulation.

As noted previously, a 4.23 s time adjustment, with no adjustment for Secretariat, is likely greater than the actual time difference between the tracks Man o' War ran compared with those on which Secretariat ran.

Therefore, it represents an unfair advantage for Man o' War. Readers, however, must ultimately decide on this conclusion's validity.

What White Stockings (or Socks) Prove

It should hold little surprise that Man o' War and Secretariat could represent the Thoroughbred equivalent to 'genetic bookends' for the Twentieth Century, or that they were probably as equivalent as two great runners, subject to racing's vagaries, might be.

Despite their distinctly varied running styles—surging to and holding the lead versus perpetually coming from far behind—they were, in fact, more directly related than many racing buffs might suspect.

All Thoroughbreds are related. This is known (Cunningham 94) just as it is now known that all humans are ultimately related. (Sykes 276)

More pertinent to this case, Fair Play (foaled 1905) sired Man o' War out of Mahubah (foaled 1910), but the same Fair Play was ultimately a paternal great-great-great grandsire of Secretariat, because he was the great grandsire of Miss Disco (foaled 1944) in direct male line through her grand sire Display (foaled 1923) and then through her sire Discovery (foaled 1931). And Miss Disco was the dam of Bold Ruler (foaled 1954), Secretariat's sire. (Mackay-Smith 171)

And so it is not difficult to see how these two superlative runners, also differing in temperament but achieving equivalent results on their respective tracks, were of similar stature (16.2 hands), girth (74-75 inches), weight ($\approx$1200 lbs.) and stride (25-26 ft.)

They both even displayed a star and a stripe. They both glistened chestnut on a sunny day. About the only significant physiognomic differences between them were Secretariat's white stockings.

Man o' War lacked stockings while Secretariat sported three—only his left front foreleg being skipped by the genetic artist.

Chapter Ten

It is also strange regarding this characteristic in that one old adage among horse aficionados is to the effect that white stockings indicate less running skill—the more stockings, the less skill.

I include this particular opinion for its relevance to the ultimate veracity of all generic opinions such as it and era differences embody.

To wit: none other than the undefeated Eclipse (foaled 1764) had one white stocking on his right rear leg; the Darley Arabian (foaled 1700) had three white stockings, à la Secretariat, except his foreleg marking reversed Secretariat's; undefeated Lexington (foaled 1850), once holder of the world's record for 4 miles, had four white stockings. (Mackay-Smith 146)

Lexington is a paternal great-great grandsire of Fair Play through Fair Play's great grand dam Aerolite (foaled 1861), his grand sire Spendthrift (foaled 1876) and his sire Hastings (foaled 1893).

And Eclipse is a paternal great-great-great grand sire of Lexington. And none other than the Darley Arabian is a paternal great-great grand sire of Eclipse—whose sire was Marske (foaled 1750), whose sire was Squirt (foaled 1732), whose sire was Bartlett's Childers (foaled 1716), who was full brother to undefeated Devonshire (Flying) Childers (foaled 1715), who set the remarkable Newmarket record—still unsurpassed—in 1721 for 3.8 miles.

Did I fail to mention, amidst all this genealogical meandering, that Flying Childers—you may, by now, have guessed—wore four white stockings? (Mackay-Smith 49)

And that, dear reader, is the end of this tale—pun definitely intended!

Chance favors only the prepared mind.
~ Louis Pasteur

Chapter Eleven

An Extraordinary Triple Crown

Man o' War did not win the Triple Crown. That was entirely because he did not run in the Kentucky Derby. As discussed in Chapter 4, Samuel Doyle Riddle, his owner, thought it was too early in the sophomore racing season of 1920 for his prize colt to carry 126 pounds for 10 furlongs.

The Triple Crown was not actually established when Man o' War raced his final year. Omaha was the first winner of a recognized, but still not formally declared, Triple Crown in 1935. He was a son of Gallant Fox, the 1930 winner of the three to-become-triple-crown races.

The formal recognition of an established Triple Crown did not occur until December 1950 at a Thoroughbred Racing Association dinner in New York. (Drager xi)

By that time eight colts had either won the triple-crown races or the Triple Crown—you have your choice.

Those colts were: Sir Barton (1919); Gallant Fox (1930); Omaha (1935); War Admiral, son of Man o' War (1937); Whirlaway (1941); Count Fleet (1943); Assault (1946) and Citation (1948).

There then occurred a twenty-five year hiatus in winners until Secretariat won in 1973.

Seattle Slew (1977) and Affirmed (1978) duplicated Secretariat's victories in all three races, minus his panache.

Since 1978 when Affirmed won, no horse has won the Triple Crown. It has thus been 35 years since the last win—ten years more than the gap separating Citation from Secretariat.

This final chapter gives an overview of Secretariat's remarkable achievements in the three races which are considered the most prestigious in North America.

Enter Two Faces of Linear Regression
Inter-Event Regression

Linear regression may be applied in two ways to predict times relative to Thoroughbred races.

The most standard application is that in which three or more separate races are used from which to derive a linear regression equation using, for example, Excel. For this chapter, I denote the more familiar type of application 'inter-event' regression. The prefix 'inter' means 'between.'

That is, a certain time passes from one race to the other and the races generally differ in location and distance. The final regression equation 'fills in,' if you will, the distance gaps which separate the races. It allows one to find a predicted time a horse is likely to run any specified distance reasonably close to the range of distances on which the regression equation was based.

For example, since this chapter focuses on Secretariat's Triple Crown, we can profitably examine the races he ran as a three-year-old leading up to the Kentucky Derby, the first of the Triple Crown events. This allows progressing through the remaining two Triple Crown events to see exactly how linear regression operates.

Chapter Eleven

Without giving full details, the distances and times in which Secretariat ran the Bay Shore, the Gotham and the Woodward can be used in a linear regression to predict his time for the Kentucky Derby.

Secretariat won the seven-furlong Bay Shore in 83.20 s. He won the eight-furlong Gotham in 93.40 s, but he lost the nine-furlong Wood Memorial by running it in 110.39 s, in third place to Sham by four lengths, with Sham trailing the winner Angle Light by a head. Angle Light's time was 109.80 s. It was Sham's final 'win' over Secretariat.

I believe it also fair to say that Angle Light ran the race of his life in the Wood Memorial.

If three successive linear regressions are derived, and updated after each run by including the latest race, the following sequences are obtained:

First—using only the Bay Shore, the Gotham and the Wood Memorial to *predict* the upcoming Kentucky Derby, Secretariat's linear regression equation is $\hat{y} = 108.76\ x - 13.10$. The COD is 0.9796, and Excel's FORECAST function predicts that he will run the Kentucky Derby in 122.85 s. He wins in a new record time of 119.40 s. The percent error between prediction and actual time is -2.81.

Second—adding the Kentucky Derby results to the first three races and running a new linear regression yields the equation $\hat{y} = 100.47\ x - 5.15$. The new COD is 0.9859.

Excel's FORECAST function predicts Secretariat will run the Preakness in 114.16 s. His actual time, after many years of delayed official reaction, was 113.00 s. The revised error between the predicted and actual times is now -1.01 percent.

Third—adding the Preakness results to the previous four races and running a third linear regression yields the prediction equation ŷ = 99.20 **x** − 4.00. Now the COD is 0.9865. When FORECAST is now run to predict the time for the upcoming Belmont, the result is 144.79 s. Secretariat ran it in 144.00, setting a new stakes, track and world record.

The error between the predicted and actual times is now -0.55 percent.

It is obvious that each time the regression is updated by more data, the equation changes and the prediction gets better—as, apparently, does Secretariat. One senses a relationship here!

This series of races provides an example of the inter-event application of linear regression.

Intra-Event Regression

It is perfectly acceptable to use linear regression solely *within* a given race. By this is meant that the times at each call point of the race, and the distances for those internal points, are used to derive the linear regression. This is the implication of the prefix 'intra,' meaning 'within,' in the process' name.

The Belmont will be the sole focus of this second exercise because, as will become evident, it probably exemplifies the greatest single performance in the history of Thoroughbred

Chapter Eleven

racing—all such statements being at least partly hyperbolic.

Secretariat's Belmont Stakes: June 9, 1973

The standard call points for the Belmont Stakes, which Secretariat ran on June 9, 1973, are one-half mile, three-quarters of a mile and one-and-one-quarter miles. The finish itself serves as the fourth and final call point.

Secretariat's times to these distances were: 46.20 s, 69.80 s, 119.00 s and 144.00 s, respectively.

Deriving a linear regression based on these four distances and times allows one to predict, and to check the prediction against reality, the exact times when Secretariat attained any given distance during the race.

The resulting intra-event linear regression for the Belmont Stakes is: $\hat{y} = 97.92 x - 3.17$. The COD is 0.9999 and the standard error of estimate is 0.51.

When Excel's FORECAST function is used with the equation to predict Secretariat's times for each furlong of the Belmont, the results are as shown in Data List 27, with times in seconds and distances in miles.

Data List 27
Furlong-by-furlong 1973 Belmont Stakes prediction

Distance	Predicted Time	Actual Time	Percent Error
0.125	9.07	12.20	-25.66
0.25	21.31	23.60	-9.70
0.375	33.55	35.00	-4.14
0.5	45.79	46.20	-0.89

Data List 27 continued

0.625	58.03	58.20	-0.29
0.75	70.27	69.80	0.67
0.875	82.51	82.00	0.62
1	94.75	94.20	0.58
1.125	106.99	106.20	0.74
1.25	119.23	119.00	0.19
1.375	131.47	131.20	0.21
1.5	143.71	144.00	-0.20
1.625	155.95	157.60	-1.05

Certain facts about Data List 27 deserve particular attention.

The percent error in the fourth column is much higher for the first three prediction distances. That is because those distances are outside the boundaries for which the original linear regression was established—namely, from 0.5 mi to 1.5 mi.

As soon as the 0.5-mile call point is reached, the error drops suddenly.

At 1.375 mi (11 f) Secretariat's actual time was faster than the time predicted by the regression equation.

The small prediction error at 1.5 mi is, again, due to the fact that 1.5 mi was the upper boundary of the data range used to establish the original regression.

Although 1.625 mi, or 13 f, was beyond the Belmont Stakes distance, Secretariat was actually timed in running out this final furlong with his jockey standing in the irons and not even hand riding him. The clocker's name was Sonny Taylor (Nack 403).

Referring to Data Lists 27 and 28 shows that, in traveling the course of the Belmont Stakes, for which Ron Turcotte said

Chapter Eleven

he was running easy throughout, Secretariat actually broke three world records. These are highlighted in the list.

Data List 28
North American Dirt Record Comparisons with the Belmont Stakes

Distance	WR at pt	Horse	Year of WR	Track
0.50	43.40	Tricky Thinker	2006	Stampede
0.625	55.20	Chinook Pass	1982	Longacres
0.75	66.49	Twin Sparks	2009	Turf Paradise
0.875	79.40	Rich Cream*	1980	Hollywood Park
1.00	92.20	Dr. Fager	1968	Arlington
1.125	105.00	Simply Majestic	1988	Golden Gate Fields
1.25	117.80	Spectacular Bid	1980	Santa Anita Park
1.375	132.31	Demi's Bret	1997	Aqueduct
1.50	144.00	Secretariat	1973	Belmont Park
1.625	158.20	Swaps	1956	Hollywood Park

*Also Time to Explode in 1982 at Hollywood Park

Since 1.1875 miles, the Preakness distance, is an important, though not an even furlong mark, point within the Belmont, it is not in Data List 28. Let it be noted, however, that Riva Ridge, Secretariat's stable mate, is co-holder of the record for that distance—at 112.40 s. His record was set in 1973 at Aqueduct. Farma Way also tied it in 1991 at Pimlico. (ARM 911)

When Secretariat's actual times from column three in Data List 27 are compared with the world-record (WR) times from column two in Data List 28, he fares extremely well, considering that he was running on his own without urging and was supposedly parceling energy for the entire 12-furlong Belmont Stakes.

At four thru thirteen furlongs, Secretariat's times compared with the world records—including those established either before his career began or as of 2011 were: +2.80, +3.00, +3.31, +2.60, +2.00, +1.20, +1.20, -1.11, 0.00 and -0.60.

All times are in seconds. The '+' indicates Secretariat's time at the given distance was higher than the 2011 world record, while the '-'indicates his time was faster than the world record. At 12 furlongs, the '0.00' indicates Secretariat's record set during the race.

The record Secretariat surpassed at 1.375 miles, or 11 furlongs, was 132.31 s set by Demi's Bret in 1997 at Aqueduct. Man o' War's previous world record for that distance, set in the 1920 Belmont Stakes, was 134.20 s. Secretariat's time at 11 furlongs in the Belmont Stakes was 131.20 s.

Signatures of Greatness

The Belmont distance changed from 11 furlongs to 12 furlongs in 1926 and has remained at that distance since. Technically, Man o' War's Belmont Stakes record on that track for that distance was never broken—as a record for an entire race. It was, so to speak, 'retired' due to the distance change in 1926.

Subtracting Secretariat's time for 11 furlongs in the Belmont Stakes from Man o' War's previous 11-furlong record gives: 134.20 – 131.20 = 3.00 s. Taken at face value, one could suggest that the Belmont track of 1920 was three seconds slower to the mile than it was in 1973 when Secretariat ran. This implies that Man o' War and Secretariat were each running at their peak efficiencies in their respective races and that *no factors other than track effect* influenced either colt's finish time.

Chapter Eleven

Although this is unlikely, it presents a strong argument that the *maximum* time adjustment Man o' War should receive for this particular distance on this particular track versus Secretariat is 3.00 s.

The 2012 ARM (1035) indicates that the Belmont Stakes 12-furlong average time for 1926 thru 1935 was 151.70 s. These are the closest years to 1920 for which the Belmont was run at 1.50 miles. For the years 1968 thru 1978, excluding Secretariat's time, the average was 149.12 s. The difference is: 151.70 − 149.12 = 2.58 s.

It is easily determined that Man o' War averaged 12.20 s per furlong for his 11-furlong run. If he is granted that same average for another furlong, a reasonable assumption, then he could have run a 12-furlong Belmont Stakes in 146.40 s, some 2.40 s slower than Secretariat's time.

Numerical guessing games could be played forever. That is not this book's purpose. However, in the interest of a more complete look at Secretariat's Belmont versus the time in which Man o' War could have run the same distance, the following values are given.

1) From Man o' War's linear regression for his 3yo season, $\hat{y} = 99.353 \mathbf{x} - 1.188$, direct substitution of 1.50 mi for $\mathbf{x}$ results in the predicted time of 147.84 s. This differs from Secretariat's record 144.00 s by 3.84 s.

If the Belmont track was truly 3.00 s slower for 11 furlongs in 1920, then it reasonably could have been proportionally slower by 3.27 s for 12 furlongs.

2) Subtracting 3.27 s from the regression result gives: 147.84 − 3.27 = 144.57 s. This is 0.57 s slower than Secretariat's 1973 world-record time.

This latter result seems fair, although not absolute, and a good place to stop approximating. In defense of this decision, if the 11-furlong times for the years 1915 thru 1925 are prorated to 12 furlongs and those times are averaged, the result is 151.01 s. This purposely excludes Man o' War's prorated time in order to err on the side of favoring the earlier runners.

The difference between this prorated average and the *actual* ten-year average, excluding Secretariat's time, for the years 1968 thru 1978 is: 151.01 − 149.12 = 1.89 s. Many readers may be surprised by this result and may notice that it conforms nicely to Walter Farley's approximation, although for four extra furlongs. (337).

That the runners from 1915 thru 1925 were possibly favored by the adjustment is evidenced by determining the average for the full 12-furlong Belmont in the years 1926 thru 1935. The value is 151.70 s. It is thus 0.69 s *slower* than the prorated 151.01 s average.

As much as demonstrating both Man o' War's and Secretariat's extraordinary running ability, the above Belmont Stakes analysis provides convincing evidence that linear regression, when applied to sufficient data, provides impressive accuracy for predicting time versus distance relations—either within a given race or between multiple races.

About Track Speed

Throughout this book various applications of Speed Ratings and Track Variants have been performed. These were basically done in seeking unbiased comparisons of Man o' War and Secretariat.

In fact, *there exists no precise scale*, in any scientific sense, on which different tracks can be placed and objectively compared.

All handicapping methods, in the final analysis, depend upon an average of some sort by which the horses they wish to compare are judged.

I addressed this issue in both GHA and BG. I shall present a slightly different view of it now.

Experienced handicapper-writers such as Ainslie (201) and Davidowitz (107) devote portions of their studies to the relativity of track descriptors such, as used by DRF, as 'fast' or 'slow.'

These time-honored terms generally do not remotely allow a researcher to make definite conclusions on a level such as: Track A is 12% slower than Track B.

The experts, in fact, stress that either Track A or Track B, or both, will likely vary in speed from day to day and even from hour to hour throughout the course of a single day's events.

Is there some reasonable way out of this undesirable situation?

I suggested one possible solution, particularly in BG, concerning a remotely controlled, robot-like device that could be guided along a racing strip for a known distance while also controlling the engine to apply constant torque to its wheels—regardless of the track's retarding forces it encountered.

If one was assured that at least near-constant torque was being applied regardless of track effects on the wheels, then the speed that the device *should have attained*, perhaps on some ideal surface such as macadam, could be calculated and compared with its actual speed.

The retardant effect of the track surface would then be accurately known.

Given the foregoing, the average speed for measured distance of this device would need to be correlated in some way with the average speed of a sample of Thoroughbreds running the same distance along the same track under conditions as nearly the same as when the track was tested. This poses a difficult task.

This procedure *might,* but not with certainty, allow an accurate determination of each horse's potential maximum speed compared with what the track permitted on that particular initial trial.

The process may sound complicated, but it is not. I know of no investigation that has been done or is currently being studied that can give the same potentially sound information regarding track speed as this technique might.

Absent this or similar equivalent studies, the problem of ambiguous terminology regarding track speed will continue.

A Related Issue

The track surface effect is, in fact, just one of two issues needing resolution. The robot just described is necessary precisely because *no living creature can duplicate the constant-torque feature demanded* for a scientific evaluation.

Even if a creature existed that was potentially capable of such muscular control, there is no way a human can communicate with him regarding what maximum speed to maintain or for how long—never mind about duplicating what was done from one trial to the next.

A small matter of language incompatibility intervenes!

Chapter Eleven

A Matter of Energy

The particular substance used for a track, whether it is a traditional loamy soil or a synthetic such as ViscoRide (42) absorbs thrust energy from the horse's hooves and legs, the hind legs in particular, through material compaction, scattering, friction and sound generation with attendant energy loss in each process.

Theoretically, the amount of energy absorbed during each stride cycle ultimately determines the runner's decrease in speed per furlong.

When track speed is mentioned it is basically implied, minus actual explanation, that different tracks cause differing amounts of speed loss per furlong, or that the same track causes such loss due to repeated use throughout a day and to changing weather.

Such qualitative statements may be true, but an actual numerical value, probably in terms of speed loss per some stated distance, is required to give real meaning to concepts such as Speed Rating and Track Variant.

As in nearly every facet of Thoroughbred data comparison, generalizations nearly always encounter trouble.

I have read comments by various students of racing that Santa Anita is probably faster than most eastern tracks. However, a compilation of the top tracks for near world-record times, taken from the 2012 ARM, shows that Belmont Park holds nearly 31% of world records set for six furlongs thru 10 furlongs between the years 1991 to 2011 and that Santa Anita was second with a bit more than 16% of such records.

Of the 75 American tracks listed in the 2012 ARM, eight accounted for slightly over 92% of the world records set during the reported twenty-year time span.

Temporary Resolution of Track Speed Problem

Perhaps the best that can presently be done, absent serious experiments on track surface effects, is to use current world records for stated distances as comparison standards.

If, for example, we wish to compare two tracks over some time period regarding how hospitable they are to eight-furlong races, we can collect a reasonably large sample, say 30 or more *best* times, over some reasonable number of days and for the same grade level of race.

A reference such as the current ARM shows that Dr. Fager holds the eight-furlong world record, set at Arlington Park in 1968, in 92.20 s. This provides an essentially absolute measure against which the subject tracks can be compared.

If the times posted on the tracks of interest were by horses of the same general grade level as Dr. Fager, then we expect that, given their best efforts, they also were capable of running eight furlongs in 92.2 s. That they did not, especially since the sample is considered large, suggests that the tracks were slower than was Arlington Park when Dr. Fager set his record.

Therefore, if the top average 30 times for each of those tracks is 92.7 s for track A and 93.1 s for track B, it is reasonably accurate to say that the absolute difference between these best times, 93.10 s – 92.70 s = 0.40 s is how much slower track B is than track A.

The statement is consistent because it is also the relative difference between tracks A and B compared to Arlington Park's 92.20 s record

That is, 93.10 – 92.20 = 0.90 and 92.70 – 92.20 = 0.50. Consequently, direct subtraction gives 0.90 – 0.50 = 0.40, the original comparison value.

Chapter Eleven

Asides and Final Thoughts

An interesting fact, based upon simplistic thinking, emerges when one compares Man o' War's nine-furlong races with the two nine-furlong races won by Secretariat. Recall that Secretariat lost two of his nine-furlong events due to documented physical problems already discussed. Thus, it is unfair to include those times in this analysis.

Man o' War's winning times were: 111.60 s at Pimlico on May 18, 1920 and 109.20 s at Aqueduct on July 10, 1920. His means and standard deviations, respectively, were: 110.40 s and 1.697 s.

Secretariat's winning times were: 107.00 s at Arlington Park on June 30, 1973 and 105.40 s at Belmont Park on September 15, 1973. His respective means and standard deviations for these races were: 106.20 s and 1.131 s.

The minus-three-sigma time for Man o' War, based on the stated data is: 110.40 − 5.091 = 105.31 s. This is his projected *fastest likely nine-furlong time* based on assuming a normal distribution for his data.

Secretariat's corresponding lower-limit time for nine furlongs is: 106.20 − 3.393 = 102.81 s.

The difference between these projected limits is 105.31 − 102.81 = 2.50 s. Since I have assumed the colts *absolutely equal in ability*, this difference must be attributed to the *track differences alone*, all other factors being equal.

This time difference reasonably corroborates all that has gone before in these analyses. In addition, it is in very general agreement with Walter Farley's statement that tracks on which Man o' War ran were *almost* two seconds slower to the mile than current tracks [implying 1962 when his book was published].

243

Since I took his word 'almost' to mean 1.90 s and based my nine-furlong proportion on that value, any discrepancy between this estimate and truth is a matter of inter-personal quibbling!

As a final note on estimation consistency, recall that the trapezoidal method detailed in Chapter 10 gave an estimated time difference for the nine-furlong races of 3.03 s.

Cardenas used an independent estimate based on similar principles for five selected sophomore races of Man o' War and four for Secretariat. His result was a mean differential of approximately 3.25 s. Cardenas (3).

I used an extended trapezoidal adjustment for all sophomore races of both colts compared with the linear regression representing the current North American Dirt records from five through 13 furlongs published in the 2012 ARM. (911)

The result was that a mean differential between linear regression lines of 2.93 s would render the total areas beneath the regression lines of both Man o' War and Secretariat exactly equal to that beneath the world-record trend line for the same range of distances.

I submit that the closeness of these disparate comparisons is significant. They are separated by only $3.25 - 2.93 = 0.32$ s. It is doubtful that any general handicapping technique, since they all rely on averages of some type and extent, can claim results for projected race times consistently closer than this.

I further submit that, since Man o' War and Secretariat represented essentially equivalent running ability, this difference

Chapter Eleven

results from a combination of track and other miscellaneous and basically imprecisely quantifiable effects.

Thus, the only conclusion is that the tracks Man o' War ran were slower than Secretariat's by somewhere within this range, a likely value being the mean, 3.07 s.

> The wind goes towards the south, and veers to the north; round and round goes the wind, and on its circuits the wind returns. All the rivers run into the sea; yet the sea is not full; to the place where the rivers flow, thither they return.
> ~ Ecclesiastes, 1: 6-7

Epilogue

Embracing the Statistical Gauntlet

Attempting to compare the past century's two greatest North American Thoroughbred runners has been a unique challenge.

Throughout the effort, striving to equitably adjust the numbers representing the souls, if you will, of Man o' War and Secretariat constantly reminded me of the stewardship charged to all humans by Israel's most illustrious prophet. (Genesis 1: 26-28)

As noted several times within the text, I was continually aware that many people do not want Man o' War compared with any other runner. Indeed, many admonitions against such comparisons have been issued over many years by his devotees.

I wish to state one final time, however, that I do not take readily to dogmatic assertions in general and to those I perceive as having thinly veiled hidden agendas, particularly.

I presented multiple reasons from the beginning why it is irrational to erect a wall—an *ad hoc* era difference—in attempting to hide what truth may be found in an honest comparison of the available racing data.

The analyses resulting from my recalcitrance strongly suggests that somewhere between two and three seconds, especially for the sophomore year, is probably the *most* time compensation which can be justly, or necessarily, attributed to Man o' War in order to compare him equitably with Secretariat. It is all that is needed.

The 'quick-and-dirty" but reasonable estimates are 1.74 s for the six-furlong races and 2.36 s for the nine-furlong races. These values are simply the differences between their average times for the respective distances. These numbers are as simple and unassuming as truth generally is.

As was also suggested, if an era difference truly exists it is comprised of a finite, though unknown, count of hoof beats or strides contained at maximum running speed within an appropriate time frame. It is no more and no less.

I also realize that, in one sense, the world's fate does not hinge on what I or any other analyst says on the matter. Nothing offered by humans can subtract one iota from the greatness of these two brilliant colts.

Intentionally, or ignorantly, contrived conclusions can only, in the final analysis, discredit the perpetrators, not the creatures.

Whether one Thoroughbred runner was, in fact, of greater or lesser ability than the other—a question I refuse to consider as proper within statistical analyses—is not truly a question of earth-shaking proportion.

What is, I believe, important to gain from such efforts is that it is humanly relevant to feel that yearnings and strivings for fairness have import. If, as a species, we lose faith in this self-assessment, then it seems that earth itself will eventually suffer.

In this regard, we may, in fact, already be ascending the final rise toward Armageddon's arid precipice.

Epilogue

The Challenge Continues

Such considerations lead me to re-evaluate my connections with Nature and Her creatures—which I have always instinctually loved. And I often like to muse that it is my inner Gaelic demanding as much.

Man o' War and Secretariat entered the flow of human and animal history during a century of severe world turbulence—an effect for which we humans appear to have seemingly inexhaustible predilection.

They did as they were 'bid,' in one sense, to survive. On another scale, judged not according to arbitrary human whims, but more consonant with a Platonic standard, Man o' War and Secretariat bequeathed lessons of the spirit 'written' in the unique shapes and patterned codes of their racing plates upon the sands and the loamy soils and the turfs of twelve different race tracks run a combined 42 times. (DRF 28, 279)

These markings now form the essence or distillate of their lives—erased forever from the sand and the loam and the turf, but transformed and archived as inky marks on past performance records which but grudgingly, it seems, yield their authors' secrets—by how deeply they inspire researchers having sensitivity enough thus to interpret them.

A time-honored admonition challenges those having eyes to see and ears to hear to attend.

Science and Beyond

Man o' War and Secretariat now rest forever beneath physical adornments atop the renowned Blue Grass which is the Kentucky Commonwealth's special and glorious emblem.

One monument is imposing, an equine bronze of noted sculptor Herbert Hazeltine. The other is somberly reflective, a modest stone by an unnamed artisan, having but that resounding five-syllable name and two hyphen-linked dates boldly carved into its grainy substance.

One stands regal and impatient, as befitting the temperament it represents; the other in simplicity is also true, subdued and seeming to wait patiently for time's proper resolution.

The first great chestnut's ironic repose is where, though born, he never raced; the second's is far westward of his Virginia home but where his surging runs induced enduring memories.

They lie now, as the proverbial crow must forever fly, exactly eleven and one-half miles apart but for a quibbling decimal. (FCC 1) Reflection suggests that 92 furlongs leaves it better said.

Their natures, as in life, cannot rest - requiring little time, setting themselves down as noble character always must, in their realm beyond human direction, to meet and visit as they choose.

Through the seasons' oft' bittersweet passing—and some 96 had come and gone since the unknown carver set his stone—observant locals eventually saw, on especially clear days, the Blue Grass seemingly 'alive' beyond its grassy nature—creating

Epilogue

gentle, silver ripples in opposing streams from just southeast of Paris to Lexington's near northeast.

Many were the explanations offered in gloried science's name, but all were ultimately discredited and reluctantly cast aside.

One day a determined, practical man and friends timed the grassy swells and found the rivulets meeting and ceasing a while, precisely at the midpoint of the interment sites of Man o' War and Secretariat, their watches reading 9:12.00.

Soon they recognized why the time—12 seconds per furlong—was an interval science could not explain and equine runners only whispered among themselves in awe.

An epiphany's gentle tremor then briefly touched them, but soon it vanished in the lowering, bright Kentucky sun.

Then smiling and casting good-byes into a sighing breeze, they departed to their homes.

For, well they knew that wind and sea—and spirits too—must always return to begin anew.

Coda

My academic training includes an MS in physics and a PhD in educational psychology. Nevertheless, this does not mean that I believe in the exclusive ability of the sciences to discover absolute truth.

I have strong leanings, in fact, toward the ideas related to intelligent design in nature rather than to a strictly "methodological materialism," espoused by most main-stream scientists.

A quotation by noted Harvard geneticist, Richard Lewontin, cited in Meyer (386), is pertinent in this regard:

> We take the side of science in spite of the patent absurdity of some of its constructs, in spite of its failure to fulfill many of its extravagant promises of health and life, in spite of the tolerance of the scientific community for unsubstantiated just-so stories, because we have a prior commitment, a commitment to materialism. It is not that the methods and institutions of science somehow compel us to accept a material explanation of the phenomenal world, but, on the contrary, that we are forced by our a priori adherence to material causes to create an apparatus of investigation and a set of concepts that produce material explanations, no matter how counter-intuitive, no matter how mystifying to the uninitiated. Moreover, that materialism is absolute, for we cannot allow a Divine Foot in the door.

Throughout this book I adhered, as closely as my temperament allowed, to the more restricted scientific pursuit of the truth about two magnificent Thoroughbred runners – as far as that truth was represented and investigated by statistical analysis.

In the Epilogue my spiritual intuitions were allowed expression. For, in my perception, truth has at least two faces – the more common names for which are matter and energy.

Absolute truth, however, no matter the direction from which it is approached, must always remain absolute. I believe that an earlier American writer who first had trouble selling his books would agree:

> *In accumulating property for ourselves or our posterity, in founding a family or a state, or acquiring fame even, we are mortal; but in dealing with truth we are immortal, and need fear no change nor accident.*
>
> ~ Henry David Thoreau
> Walden

References

Ainslie, Tom. Ainslie's Complete Guide to Thoroughbred Racing. 3rd ed. New York: Fireside-Simon & Schuster, 1988.

All Breed Database. "Herod (horse)." 2013. 1 Oct. 2013 http://www.allbreedpedigree.com/herod.

---, "Matchem (horse)." 2013. 1 Oct. 2013 http://www.allbreedpedigree.com/matchem.

Anderson, Abigail. "A LIVING FLAME: WILL HARBUT AND MAN O' WAR." The Vault. 23 August 2011. 8 Apr. 2013 http://thevaulthorseracing.wordpress.com.

Anderson, Elmer E. Modern Physics and Quantum Mechanics. Philadelphia:W. B. Saunders Company, 1971.

Associated Press. "Secretariat's Winning Time in '73 Preakness Changed." The Daily Record 19 Jun. 2012. 15 Jul. 2013 http://thedailyrecord.com/2012/06/19/secretariats-winning-time-in-73-preakness-changed/#ixzz2PmdZKZbv.

"Astronomical Constants." Time Almanac. 2004.

Beyer, Andrew. Beyer on Speed: New Strategies in Racetrack Betting. Boston: Houghton Mifflin Company, 2007.

Campbell, Jed. "Why is 30 the "Magic Number" for Sample Size?" 18 Apr. 2011. 10 Jan. 2013 http://www.jedcampbell.com/?p=262.

Cardenas, Russell A. "Secretariat and Man o' War Revisited." 2006: 1-24. 5 Oct. 2013 http://www.truevine.net/~sec@truevine.net/Legends/.

Colin's Ghost.org: "Colin: One of the Great Ones." 4 Jun 2010. 8 Oct. 2013 http://colinsghost.org/2010/.06/colin-one-of-the-great-ones.html.

Cunningham, Patrick. "The Genetics of Thoroughbred Horses." Scientific American 264 (1991): 92-98.

Daily Racing Form. "Help: How to Use DRF – Speed Figures." 23 Jun.2013 http://www1.drf.com/help/help_speedrate.html.

---, Champions: The Lives, Times, and Past Performances of the 20th Century's Greatest Thoroughbreds. Rev. ed. New York: Daily Racing Form Press, 2005.

---, The American Racing Manual. Ed. Paula Welsh Prather. New York: Daily Racing Form Press, 2012.

Davidowitz, Steven. Betting Thoroughbreds: A Professional's Guide for the Horseplayer. New York: E.P. Dutton, 1977.

Denny, Mark, and Gaines, Steven. Chance in Biology: Using Probability to Explore Nature. Princeton: Princeton University Press, 2000.

Downing, Douglas, and Clark, Jeffrey. Baron's E-Z Statistics. Hauppauge: Baron's Educational Series, Inc., 2009.

Drager, Marvin. The Most Glorious Crown: The Story of America's Triple Crown Thoroughbreds from Sir Barton to Affirmed. Chicago: Triumph Books, 2005.

ESPN Enterprises, Inc. The Life and Times of Secretariat: An American Racing Legend. Dist. Capital Cities/ABC Video Publishing, Inc., 1990.

Farley, Walter. Man o' War. New York: Yearling-Random House, 1962

Federal Communications Commission. "Find Distance and Azimuth between 2 Sets of Coordinates." Dec. 1998. 12 Sep. 2013 http://transition.fcc.gov/fcc~bin/distance?

Ferguson, George A. Statistical Analysis in Psychology and Education. 4th ed. New York: McGraw-Hill Book Company, 1976.

References

Google. "Arlington Park Racetrack – Sightseeing with Google Satellite Maps." 2 Oct. 2013 http://www.satellite-sightseer.com/id/7433/United States/Illinois/Arlington Heights/Arlington.

Harvey, Greg. Excel all-in-one Desk Reference for Dummies. Hoboken: Wiley Publishing, Inc., 2007.

Hollingsworth, Kent. The Kentucky Thoroughbred. Lexington: The University Press of Kentucky, 1985.

Justice, Charles. The Greatest Horse of All: A Controversy Examined. Bloomington: Author House, 2008.

---, Beyond Greatness: Four Thoroughbred Legends. Bloomington: Author House, 2011.

Mackay-Smith, Alexander. Speed and the Thoroughbred: The Complete History. Lanham: The Derrydale Press, 2000.

Meyer, Stephen C. Darwin's Doubt: The Explosive Origin of Animal Life and the Case for Intelligent Design. Seattle: Harper One-Harper Collins, 2013.

McCarthy, Clem. "Blue Grass." Saturday Evening Post 13 Sept. 1941: 18-104.

Minetti, A. E., et al. "The Relationship between Mechanical Work and Energy Expenditure of Locomotion in Horses." The Journal of Experimental Biology 202 (1999): 2329-2338.

Nack, William R. Foreward. Thoroughbred Champions: Top 100 Racehorses of the 20th Century. Lexington: The Blood-Horse, Inc., 1999. 6-8.

---, Secretariat. Hyperion-Harper Collins: New York, 2010.

Ours, Dorothy. Man o' War: A Legend Like Lightning. New York: St. Martin's Press, 2006.

Robinson, Tara Rodden. Genetics for DUMMIES. Hoboken: Wiley, 2005.

Schmuller, Joseph. Statistical Analysis with Excel for DUMMIES. Hoboken: Wiley Publishing, Inc., 2005.

Sears, Francis W., Zemansky, Mark W., and Young, Hugh D. University Physics. 6th ed. Reading: Addison-Wesley Publishing Company, 1982.

Stretchrun. "Secretariat didn't reproduce himself, but sired many champions and was a great broodmare sire!" 2 Feb. 2010. 16 May 2013 http://www.horseracegame.com/community/content/blogs/stretchrun/02-02-2010/secretar ...

Sykes, Bryan. The Seven Daughters of Eve. New York: W. W. Norton & Company, 2001.

Tweedy, Kate Chenery, and Ladin, Leeanne. Secretariat's Meadow: The Land · The Family · The Legend. Manakin-Sabot: Dementi Milestone Publishing, Inc., 2011.

ViscoRide. "Track Surface Properties: Hardness." 2008. 23 Jan. 2013 http://www.viscoride.com.au/surface.html.

Walpole, Ronald E., and Myers, Raymond H. Probability and Statistics for Engineers and Scientists. 3rd ed. New York: Macmillan Publishing Company, 1985.

Wikipedia. "Eightfold Way (physics)." 23 Oct. 2013. 20 Nov. 2013. http://en.wikipedia.org/wiki/Eightfold_Way_(physics).

---, "History of Watches." 14 Dec. 2013. 15 Dec. 2013 http://en.wikipedia.org/wiki/History_of_watches.

---, "Terlingua (horse)." 13 Nov. 2013. 20 Nov. 2013 http://en.wikipedia.org/wiki/Terlingua_(horse).

Wolfe, Raymond G., Jr. Secretariat. 25th Triple Crown Anniversary Edition, 1981.

References

Ziccardi, M. James. Medieval Philosophy: A Practical Guide to William of Ockham. N.p.: 2011.

Appendix A

Basic Data: Man o' War and Secretariat

Man o' War, 1919: 2yo

Result	Track/Spd/Date	Distance	Time	SR-TV	TR	WR	Impost
win	Bel.fst.13.09	0.75	71.60	85-21	68.60	66.49	127
win	Sar.sl.30.08	0.75	73.00	87-14	70.40	66.49	130
win	Sar.fst.23.08	0.75	72.00	92-08	70.40	66.49	130
place	Sar.fst.13.08	0.75	71.27	95-09	70.20	66.49	130
win	Sar.fst.02.08	0.75	72.40	90-13	70.40	66.49	130
win	Aqu.fst.05.07	0.75	73.00	90-12	71.00	66.49	130
win	Aqu.fst.23.06	0.625	61.60	83-12	58.20	55.20	130
win	Jam.gd.21.06	0.6875	66.60	92-09	65.00	61.03	120
win	Bel.sl.09.06	0.6875	65.60	91-17	63.80	61.03	115
win	Bel.fst.06.06	0.625	59.00	83-18	55.60	55.20	115

Man o' War: Linear Regression 2yo

95.6050	**0.4884**
5.7011	4.0720
0.9723	**0.9014**
281.2185	8.0000
228.5079	6.5005

Secretariat, 1972: 2yo

Result	Track/Spd/Date	Distance	Times	SR-TV	TR	WR	Impost
win	GS.fst.18.11	1.0625	104.40	83-23	101.00	98.40	122
win	Lrl.sly.28.10	1.0625	102.80	99-14	102.60	98.40	122
win	Bel.fst.14.10	1.00	95.00	97-12	94.40	92.20	122
win	Bel.fst.16.09	0.8125	76.40	98-09	76.00	73.00	122
win	Sar.fst.26.08	0.8125	76.20	97-12	75.60	73.00	121
win	Sar.fst.16.08	0.75	70.00	96-14	69.20	66.49	121
win	Sar.fst.31.07	0.75	70.80	92-13	69.20	66.49	118
win	Aqu.fst.15.07	0.75	70.60	90-14	68.60	66.49	113
fourth	Sar.fst.04.07	0.6875	65.18	87-11	62.58	61.03	113

Secretariat Linear Regression 2yo

103.4376	**-7.0885**
2.3940	2.0715
0.9963	**0.9925**
1866.8650	7.0000
1838.9477	6.8953

Man o'War, 1920: 3yo

Result	Track/Spd/Date	Distance	Times	SR-TV	TR	WR	Impost
win	Knw.fst.12.10	1.25	123.00	132-03	129.40	117.80	120
win	HdG.fst.18.09	1.0625	104.80	101-16	105.00	98.40	138
win	Bel.fst.11.09	1.50	148.8	117-02	152.20	144.00	118
win	Bel.st.04.09	1.625	160.8	134-00	167.60	158.20	126
win	Sar.fst.21.08	1.25	121.8	102-08	122.20	117.80	129
win	Sar.fst.7.08	1.1875	116.6	97-09	116.00	112.40	131
win	Aqu.fst.10.07	1.125	109.2	101-08	109.40	105.00	126
win	Jam.gd.22.06	1.00	101.6	86-13	98.80	92.20	135
win	Bel.fst.12.06	1.375	134.2	116-10	137.40	132.31	126
win	Bel.fst.29.05	1.00	95.8	104-10	96.60	92.20	118
win	Pim.fst.18.05	1.125	111.6	97-10	111.00	105.00	126

Man o' War: Linear Regression 3yo

99.3530	**-1.1878**
2.5929	3.2212
0.9939	**1.6570**
1468.2150	9.0000
4031.1368	24.7104

Secretariat, 1973: 3yo

Result	Track/Spd/Date	Distance	Finish	SR-TV	TR	WR	Impost
win	WO.fm.28.10	1.625*	161.80	96-04	161.00	157.00	117
win	Bel.fm.08.10	1.50*	144.80	103-01	145.40	142.80	121
place	Bel.sly.29.09	1.50	146.47	86-15	143.00	144.00	119
win	Bel.fst.15.09	1.125	105.40	104-07	106.20	105.00	124
place	Sar.fst.04.08	1.125	109.35	94-15	108.00	105.00	119
win	AP.fst.30.06	1.125	107.00	99-17	106.80	105.00	126
win	Bel.fst.09.06	1.50	144.00	113-05	146.60	144.00	126
win	Pim.fst.19.05	1.1875	113.00	105-13	114.00	112.40	126
win	CD.fst.05.05	1.25	119.40	103-10	120.00	117.80	126
show	Aqu.fst.21.04	1.125	110.39	83-17	106.40	105.00	126
win	Aqu.fst.07.04	1.00	93.40	100-08	93.40	92.20	126
win	Aqu.sly.17.03	0.875	83.20	85-17	80.20	79.40	126

*Turf

Secretariat: Linear Regression 3yo

102.7080	**-8.0000**
2.4563	3.1063
0.9943	**1.8984**
1748.4768	10.0000
6301.1980	36.0382

Appendix A

World Dirt Records: 2011

Miles	Furlongs	Finish
0.6250	5.0	55.20
0.6875	5.5	61.03
0.7500	6.0	66.49
0.8125	6.5	73.00
0.8750	7.0	79.40
0.9375	7.5	86.26
1.0000	8.0	92.20
1.0625	8.5	98.40
1.1250	9.0	105.00
1.1875	9.5	112.40
1.2500	10.0	117.80
1.3750	11.0	132.31
1.4375	11.5	143.00
1.5000	12.0	144.00
1.5625	12.5	155.77
1.6250	13.0	158.20

Track Abbreviations

AP	Arlington Park
Aqu	Aqueduct
Bel	Belmont
CD	Churchill Downs
GS	Garden State Park
HdG	Havre de Grace
Jam	Jamaica
Knw	Kenilworth
Lrl	Laurel
Pim	Pimlico
Sar	Saratoga
WO	Woodbine

Appendix B
Calculating the Standard Deviation and Variance

For the set of ten data items given in Chapter 2, {8, 10, 9, 9, 7, 9, 11, 12, 11, 10}, the standard deviation is calculated using the following six steps.

1. Find the mean, $\bar{x}$, of the ten items. The mean is 9.60 from Chapter 2.
2. Subtract $\bar{x}$ from *each item* in the set. This gives the respective values: -1.6, 0.4, -0.6, -0.6, -2.6, -0.6, 1.4, 2.4, 1.4, 0.4. Place these in a column, as shown below. These values are called *deviation scores*.
3. Square *each* deviation score and arrange the results as in the second column.
4. Add the ten squared values in column 2. The result is 20.40
5. Divide 20.40 by n -1, or 9. The result is 20.40 ÷ 9 = 2.27 rounded to two decimal places. This is the *variance* of the original set of ten numbers.
6. Take the square root of 2.27. The answer, the *standard deviation*, is 1.51 to two decimal places. The standard deviation is *always* the square root of the variance.

Deviation Scores	Squared Deviation Scores
-1.6	2.56
0.4	0.16
-0.6	0.36
-0.6	0.36
-2.6	6.76
-0.6	0.36
1.4	1.96
2.4	5.76
1.4	1.96
0.4	0.16
	Total: 20.40

Total divided by n – 1, the sample size minus 1: 20.40 ÷ 9 = 2.27

The square root of 2.27 = √2.27 = 1.51 = standard deviation of original set. These steps are automatically performed by Excel when the STDEV operation is applied to an array of spreadsheet data.

The standard deviation's significance is explained in the text. Remember, the variance is the square of the standard deviation and is always a positive number.

Appendix C
Calculating the Pearson Product-Moment Correlation Coefficient r

The formula for calculating the correlation, r, between two sets of values is:

$$r = [(1/(N-1)) \cdot \Sigma(x_i - \bar{x})(y_i - \bar{y})]/s_x s_y$$

In this equation, all calculations of quantities within the brackets, [ ], are done first. That result is then divided by the product of the two standard deviations, s_x and s_y. These are the standard deviations of the sets X and Y having N values each that are, respectively, plotted along the x and y axes when the correlation is graphed.

The items in this equation are: N, the sample size; x_i, each separate value of set X, where i goes from 1 to N; y_i, each separate value of set Y just as for x; $\bar{x}$ = the mean of the N values of the x_i variables; $\bar{y}$ = the mean of the N values of the y_i variables.

For example, if you wish the correlation between the two sets of data given by X = {1, 3, 5, 7, 9} and Y = {2, 4, 6, 8, 10} then N = 5 because two sets of 5 numbers each are being correlated. Therefore, i goes from 1 to 5. Excel gives the means of sets X and Y as 5 and 6, respectively. For clarification, $x_1 = 1$, $x_2 = 3$, etc., and similarly for the y_i values.

Enter two *adjacent* columns of data in an Excel spreadsheet. Place the values from set X in a separate cell for each value, in the *leftmost* column, and place the values from set Y, each in its own separate cell, in a column *adjacent to and to the right of* the X column values. Excel only accepts data for two variables in this format.

Use Excel, as explained in the text, to calculate the means and standard deviations of both the X and Y sets. These values are $\bar{x} = 5$, $\bar{y} = 6$, $s_x = s_y = 3.16$. Then set up five columns of data as below to complete the equation for the Pearson correlation coefficient.

X	Y	$(x_i - \bar{x})$	$(y_i - \bar{y})$	$(x_i - \bar{x}) \cdot (y_i - \bar{y})$
1	2	-4.00	-4.00	16.00
3	4	-2.00	-2.00	4.00
5	6	0.00	0.00	0.00
7	8	2.00	2.00	4.00
9	10	4.00	4.00	16.00

Note also that N − 1 = 4 and the sum of $(x_i - \bar{x}) \cdot (y_i - \bar{y})$ is 40.00. All values are rounded to two decimal places. Your answer should be: r = 1.00.

Appendix D
The Output Data from LINEST Linear Regression

Basic Linear Regression Procedure

To use Excel's LINEST program correctly, data must first be entered pair-wise in two adjacent columns of a spreadsheet. The *independent variable* values must be in the *leftmost* column, and the *dependent* variable values must be in the *rightmost* column.

After the data are thus entered, you must click and drag the mouse cursor over a nearby array of cells which is two columns wide and five rows deep. No other combination will work properly with this routine. Excel's results go in this array.

Then select the 'f_x' function button and proceed to work through the drop-down menus that appear. First select 'LINEST' from the 'Insert Function' menu and then insert, in succession, the y values, x values, 'true,' and 'true' in the *four entry fields* of the 'Function Arguments' menu.

Press 'Shift-Ctrl-Enter.' The results will appear in the pre-selected, ten-cell array you first highlighted. Do not press the 'OK' button because you will not get the complete linear regression results. This is one of Excel's quirks.

The following table is an example relating to Figure 3 in the text. Each of Excel's ten output values is explained below. You actually need use only values 1, 2, 5 and 6 (shaded) from the table to make immediate practical use of these numbers.

1	Slope of regression line	2	Intercept of regression line
3	Standard error of slope	4	Standard error of intercept
5	Value of R^2	6	Standard error of estimate
7	Value of F	8	Degrees of freedom, df
9	Sum of regression squares	10	Sum of residual squares

1. The slope of the regression line was explained in Chapter 2. It is the amount of increase or decrease in the dependent variable, y, for each *unit increase* in the independent variable, x. For Figure 3 this value is: 0.394994.
2. The intercept of the regression line, also explained in Chapter 2, is the point on the y axis where the regression line would cross it if the line were extended that far backward. It is a kind of anchor point used to define the line's location. For Figure 3 the value is 2.586081.
3. The *standard error of the slope* is the *standard deviation* of the *average value* you would obtain for the slope if Excel's LINEST routine were run on many different but related samples of data. For Figure 3 the value is 0.063259.
4. The *standard error of the intercept* is the *standard deviation* of the *average value* for the intercept you would obtain if Excel's LINEST routine were run on many different but related samples of data. For Figure 3 the value is 0.910144.
5. The *coefficient of determination*, symbolized by R^2, was explained in Chapter 2 as the percent of change in the dependent variable, y, explainable by changes in the independent variable, x. For Figure 3 the value is 0.829744.
6. The *standard error of estimate* is the *standard deviation* you would expect for the *mean of a given y value* in the entire *population* of samples of related data if many samples were taken from that population. For Figure 3 the value is 1.144978.

7. The value F is the *ratio* of the values for table items 9 and 10 divided by their respective degrees of freedom, 1 and 8. It is symbolized F = MSreg/MSres. It is used to test whether the regression line significantly predicts changes in the dependent variable. For Figure 3 the value is 38.98797 which is significant.
8. The residual degrees of freedom is always 2 less than the sample size, 10 in this case, of the data entered into Excel's LINEST routine. It has this value because two data points are required, and are thus unavailable for further use, to determine placement of the regression line. For Figure 3 the value is 8.
9. The sum of squares due to regression, SSreg, is the sum of the squared values of $\hat{y} - \bar{y}$. Figure 3 in the text shows typical placement of these values. For Figure 3 the value is 51.11221.
10. The sum of squares of the residuals, SSres, is the sum of the squared values of $y - \hat{y}$. Figure 3 in the text shows typical placement of these values. For Figure 3 the value is 10.48779.

Linear Regression, Correlation and Causality

There seems to be a natural tendency among students to assume that the independent variable causes, in some sense, the resulting values of the dependent variable, especially when a high correlation exists between the two variables.

This may sometimes be the case, but it generally needs more verification than just the results of a high COD. That is, just because the COD may be 0.99 or above, as in most regressions of time against distance for Thoroughbred racing, does not mean that the distance run (independent variable) by a horse *causes* the resulting time (dependent variable) in which the horse runs that distance.

Certainly one might expect that, the longer the distance which a given horse runs, the longer it takes to run that distance. However, in the sense that quantity of rainfall *directly influences* or otherwise causes rate of grass growth, distance per se has no analogous physical or chemical effect on time – as in a horse race.

The question needs close study and interpretation in each separate case of two variables being related to each other via simple linear regression – or even in the case of multiple regression.

This caveat is presented to make readers more aware that interpretations must be approached with caution. Suffice it to say that questions of ultimate causality are philosophical and are not within the purview of statistics alone.

Significance of the Slope in Linear Regression.

The slope given by a linear regression procedure is an important part of the regressions' interpretation.

For instance, when linear regressions are run for both Man o' War and Secretariat for their sophomore years, it is found that Man o' War's regression is $\hat{y} = 99.35 \mathbf{x} - 1.18$ while Secretariat's is $\hat{y}$ $102.71 \mathbf{x} - 8.00$.

Man o' War's slope, the coefficient of $\mathbf{x}$ in the regression equation, is 99.35. It implies that, for each one mile increase in distance run, it typically required 99.35 s for Man o' War to run it.

Appendix D

Similarly, it typically required Secretariat 102.71 s to run one additional mile – at lease for the range of the independent variable used to establish his regression equation.

Thus, directly subtracting Man o' War's slope from Secretariat's gives 102.71 – 99.35 = 3.36 s.

On the surface, this appears to indicate that Man o' War could run one mile, starting at any given distance, in 3.36 s less than Secretariat.

However, directly substituting 1.00 (one mile) for x into both equations gives the $\hat{y}$ values of 98.16 s for Man o' War and 94.71 s for Secretariat. This result disagrees with the interpretation based on slope. Is something wrong with the equations?

The answer is that nothing is wrong except, perhaps, overenthusiastic interpretation.

Remember that each equation has an intercept value that must be subtracted from the result of multiplying x by the slope of the line.

In addition, there is the matter of how *significantly different from zero* is each slope number. This can be tested by dividing item #1 from the regression output table by item #3, the standard error of estimate for the slope.

For Man o' War the result is 99.35 ÷ 2.59 = 38.36. For Secretariat, the result is 102.71 ÷ 2.46 = 41.75. The difference between these two results is 3.39, and that essentially explains, although it is not the entire story, the original 3.36 s difference.

A clear discussion is given (Schmuller 231) of how the t-test for the slope, equivalent to dividing items # 1 and 3 from the LINEST output table, is performed and what its implications are.

Appendix E
Reference Table for Selected t Values of a One-Tailed Distribution*

df	α levels:	0.10	0.05	0.025	0.01	0.005
1		3.078	6.314 →	12.706	31.821	63.657
2		1.886	2.920	4.303	6.965	9.925
3		1.638	2.353	3.182	4.541	5.841
4		1.533	2.132	2.776	3.747	4.604
5		1.476	2.015	2.571	3.365	4.032
10		1.372	1.812	2.228	2.764	3.169
20		1.325	1.725	2.086	2.528	2.845
30		1.310	1.697	2.042	2.457	2.750
40		1.303	1.684	2.021	2.423	2.704
50		1.299	1.676	2.009	2.403	2.678

*Table modeled after Table B in Ferguson (487)

The highlighted items of the above tabulation show degrees of freedom as 'df' along the left border while the respective five alpha levels indicate the corresponding critical fractional area under a normal distribution. Thus, 0.10 means both $\alpha = 0.10$ and the upper one-tenth (ten percent) of the area under the distribution; 0.05 means $\alpha = 0.05$ and the upper five hundredths (five percent) of the area under the distribution, and so forth. These areas represent the right tail of the distribution, marked 'α,' in Figure E.

Figure E
Area under t Distribution Representing Probability Value of TTEST

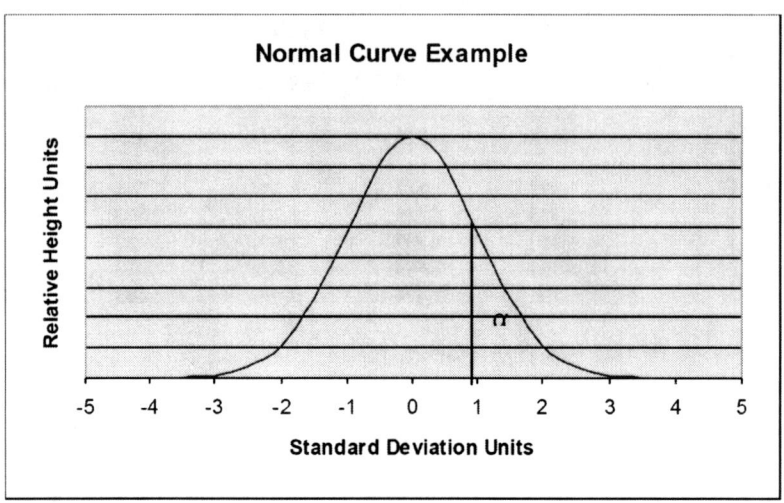

For example, if two samples each having a total df equaling 10 are being compared, the value of t at the lower boundary (nearest the distribution's mean value) of the α region containing 0.05 of the total area under the distribution is at the point t = 1.812.

The total degrees of freedom for a two-sample t-test is $(n_1 - 1) + (n_2 - 1)$, with n_1 being the number of items in the first sample and n_2 being the number of items in the second sample. The samples need not be equal size.

Use the one-tailed values given above *when you know or strongly suspect* that the values from one sample are either larger or smaller than those of the second sample. If this is not obvious by inspection, use the two-tailed test.

A two-tailed test is always used when you're uncertain of the direction of the difference in relative size between the two

Appendix E

samples. This is the case when both larger and smaller values are intermingled between samples.

For illustration, the t value 3.078 for one degree of freedom, $df = 1$, at $\alpha = 0.10$ implies that the area under the curve to the left of t is 1.00 - 0.10 or 0.90 of the total area.

This assumes a one-tailed test, as the table and Figure E indicate. Remember that the table value, $\alpha = 0.10$, defines the critical region. The remaining area under the normal distribution is 1.00 minus the critical region for a one-tailed test since the total area under the distribution is always defined as 1.00.

The interpretation of this particular example is that the probability is 0.90 or 90 % that the two samples tested could have come from populations having equal means just from random sampling *if their t-value is less than* 3.078. This would *support* a null hypothesis that the difference between the population means was zero, even though it is not exactly zero. Therefore, *the null hypothesis would be accepted in this case* and the samples would be said *not to differ significantly*.

Figure E was designed to show a distribution in standard normal form. That is, having mean 0 and standard deviation 1.00. From the discussion of normal-curve areas in Chapter 2, you may recall that 99.72% of the area under a normal curve lies between -3 and +3 standard deviations on either side of the mean. In Figure E t-values spaced 1 SD apart divide the curve into the same proportions as the standard deviation discussed in Chapter 2.

If the chosen alpha level for the *one-tailed* t-test was 0.05 and the alternate hypothesis $H_1: \mu_1 - \mu_2 > 0$ was selected, the table indicates that t must be 6.314 or greater with $df = 1$ before the lower boundary (critical value) of the distribution's upper

0.05 fractional area is reached. For this t-value or beyond the null hypothesis is *rejected*.

For an *equivalent two-tailed test, α = 0.05 must be apportioned equally into each of two tails*. The resultant probability applying to *each* tail of the distribution is then α/2 = 0.025 of the total area. Thus, the null hypothesis will be rejected if a t value of -12.706 or less occurs for the left tail or a t value of +12.706 or greater occurs for the right tail. This maintains a *total rejection area* of 0.05, the sum of the areas in the two tails.

The horizontal arrow pointing from 6.314 to 12.706 in the table shows the general direction to read – throughout the table values – to switch **from** a one-tailed **to** a two-tailed test. This method works for any two values *from adjacent columns* in the table.

When performing a two-tailed test at α = 0.05, it means that 0.95 or 95% of the total area remains under the distribution's curve and *between the limits* of the t-test values. This is the 'acceptance' region.

In this case, still assuming that df = 1, it can either mean that it is 95 percent likely that random samples with the given df and their particular sample values will have a t-value *between* -12.706 and +12.706 or that there is a 5 % chance (2.5 % in each tail) that random samples will have a t-value *within either* of the tails. These are the only possibilities.

The final entry in the TTEST Function Arguments dialog box 'asks' for equal or unequal variance between the two samples. The F Test is used to check this statistic, as mentioned in the text.

Appendix E

A '2' or '3' is entered for equal or unequal variance, respectively.

To perform the test, select FTEST by using the f_x button, just as for the TTEST. A Function Arguments dialog box appears requiring two arrays to be filled. Enter the range of cells, or click and drag them, into Array 1 for the data from the first sample, and then do the same in the Array 2 field for the data of the second sample.

The FTEST automatically performs a one-tailed test because it is measuring the ratio between two variances, with the larger of the variances always being placed in the numerator. Since variances are always positive, there is just one direction the test can go, and so the one-tailed version is applied.

If the value of F given by the test is greater than 0.05, this indicates no significant difference between the variances. You should then enter '2' in the final array field of the Function Arguments dialog box for the TTEST.

Practice using both the TTEST and FTEST in combination on data examples you make up. It will not take many trials before you feel comfortable using these tests.

Appendix F provides more explanation about the possible combinations of hypotheses that can be used in the t test for the difference between sample means.

Note that, in going from a one-tailed value of t to a corresponding two-tailed value, the magnitude of t always *increases*.

This is because each of the two-tailed rejection regions is

further from the mean of their distribution than is the single one-tailed rejection region.

The increase in distance from the mean for the two-tailed regions provides one reason for stating that a one-tailed test is generally preferable because it is more sensitive to significant differences between means than is the two-tailed test.

The one-tailed test is more sensitive precisely because one need not have as large a value of t to reach the critical value (lower boundary) of the one-tailed region as one does to reach either of the two-tailed regions.

For example, referring to the table at the beginning of this appendix, if one were testing two samples for significant difference between means at the one-tailed significance level of 0.05 with df = 10, t would need only be 1.812 or greater for the result to be significant.

If one were testing the same samples at the equivalent two-tailed level, the value of t needed for significance would be 2.228 or greater.

Some authors point out this sensitivity difference between the two types of tests, but they do not always explain the reason in the same manner. (Schmuller 148)

Appendix F
The Possible Combinations of Null and Alternate Hypotheses

General Information about t-Test

When TTEST is run in Excel, the result gives the *probability* of having a difference between sample means *equal to or greater than* the one obtained simply by random sampling if the samples are from equivalent populations.

This probability, for a one-tailed test, is *the same as* the area under a t distribution (essentially a normal distribution if the sample size is 30 or greater) beginning *at the t value Excel calculates* and extending through the remainder of the distribution's tail. Figure F, identical to Figure E in Appendix E, shows the area, labeled 'α,' for $H_1: \mu_1 - \mu_2 > 0$. It indicates both the test's significance level and the corresponding fractional area under the distribution. The value $\alpha = 0.05$ is typically accepted by statisticians for most applications.

Figure F
Showing the t-Test Probability Region α

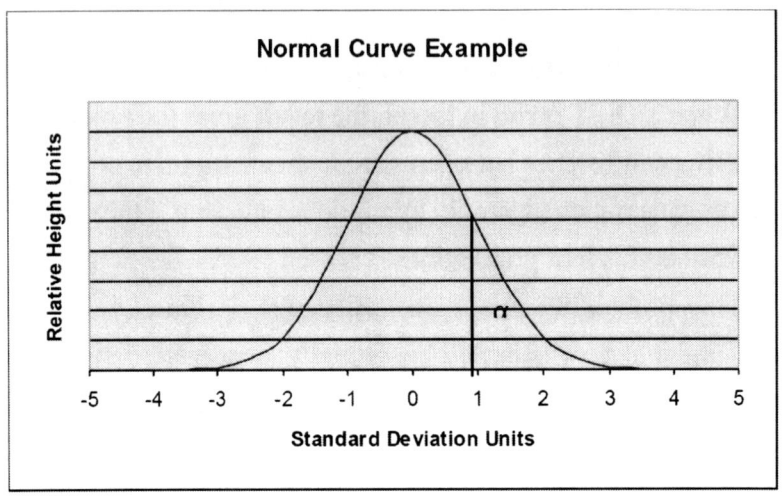

Excel *does not give* the t value to which the critical probability applies. It gives the probability only. The TINV function, however, displays the corresponding t value.

For example, if two samples comprised of the sets A = {2, 5, 7, 6} and B = {1, 3, 8, 7}, are compared using TTEST with df = 6 and a one-tailed test is specified, the probability value α = 0.451674 (0.45 rounded) is obtained. Recall that, for a two-sample t-test, df equals *two less than the sum* of the items in both samples. In this case that is 8 – 2 = 6.

This TTEST result represents *both* the *probability* that the samples could represent equivalent populations (populations with equal means) *and* the fractional *area* t cuts off in the upper portion of its distribution. Since the probability value for these samples is greater than 0.05, the *null hypothesis should be accepted*. In other words, the *difference between these sample means was not unusual under the test assumptions*.

Appendix F

TTEST automatically assumes probability 0.05 is intended unless otherwise directed.

If the *actual value of t* giving this result, *for one tail*, is desired, the f_x function button is pressed and TINV is selected from the 'Insert Function' menu. A dialog box, 'Function Arguments,' opens. It contains two fields needing completion. In the top field labeled 'Probability,' *twice* the α value (using Excel 2003) obtained in the TTEST (2 • 0.45 = 0.90) should be entered. The df (6 in this case) should be entered in the bottom field.

The reason for using twice the one-tailed probability is that TINV always runs a *two-tailed* test no matter what probability value, e.g. 0.05, is entered. This is just how Excel 2003 was designed. Entering twice the desired probability value, 2 · 0.05 = 0.10 for this example, results in the correct *one-tailed* t value being returned.

When 0.90 and df = 6 are entered in the two TINV fields, the one-tailed value t = 0.131076 is obtained. Since a one-tailed test was specified, this is the actual t value which is the lower boundary of the upper 0.45 (or 45 percent) of the area under the distribution.

For a slightly different explanation of the TINV function, see Schmuller (148).

When in doubt about number of tails for TTEST

When Excel's TTEST is run it may not be obvious at first how many tails to specify. The following discussion should clarify the selection process.

Try to determine from visual inspection whether one sample's mean is greater or less than the second sample's or whether most of the values in one sample are either larger or smaller than those in the other. If that is not possible, specify a *two-tailed* test, also called a 'non-directional' test.

For a two-tailed test, *always* assume the null hypothesis H_0: $\mu_1 - \mu_2 = 0$ plus the alternate hypothesis having the form H_1: $\mu_1 - \mu_2 \neq 0$. Anytime the symbol '$\neq$' appears in the alternate hypothesis H_1, it *signals a two-tailed test* because it implies that the investigator is uncertain whether one population's mean is less than or equal to, $\leq$, or greater than or equal to, $\geq$, the other's.

When you *can* reasonably tell by inspection or logical reasoning that one sample mean is probably greater than or less than the other, use the appropriate corresponding alternate hypothesis: either H_1: $\mu_1 - \mu_2 \leq 0$ if you think the mean of population 2 is *greater than or equal to* that of population 1 based on the evidence from their respective samples, or H_2: $\mu_1 - \mu_2 \geq 0$ if you think the opposite case holds.

If the mean of population 2 *is greater than* that of population 1, then subtracting the means in the given order will produce a negative result, and that is what '<' in the composite symbol '≤ 0' implies because it means a number *less than* 0 which is, by definition, *negative*.

Conversely, if it appears that the mean of population 1 is greater than that of population 2, you then expect the given difference to be either 0 or positive, and that obviously implies that $\mu_1 - \mu_2 \geq 0$.

Closing Thoughts about t-Tests

If the difference between the sample means actually does equal zero, then the probability that the null hypothesis should be accepted will be 0.50 or within a reasonable range thereof. This is the *maximum attainable probability* for the test. It will be

Appendix F

instructive for you to use Excel's RNG function and investigate this concept.

For example, when two *known normal distributions*, both having mean 10 and standard deviation 2 (thus identical) and containing 100 values each, were generated using the RNG and then checked with the t-test, the 10 probability values in the following list were obtained. These values represent the likelihood of their coming from equivalent populations. The mean value, shown as the last entry in the listing, was 0.29 rounded to two decimal places.

At least half *these t values appear too low* since we know, *a priori* in this case, that the means *definitely came from equal populations* as generated by Excel's RNG. In fact, the first list value, 0.033418, indicates, since it is < 0.05, a significant difference – when there should not be – between the sample means.

This provides an excellent example of how a Type I or alpha error can occur. It is a Type I error precisely because it would lead an investigator who did not know the origin of the two samples (that they were equivalent) to conclude that the null hypothesis should be rejected when, in fact, it should have been accepted.

The shaded list values are more expected and reasonable for this case. The overall result of these simulations emphasizes that nothing in known with certainty in statistics – much as the Uncertainty Principle, in a slightly different context, applies in physics.

Table F
Ten Alpha Significance-Level Values from Two Equal RNG Distributions

10 α values
0.033418
0.201734
0.266508
0.100057
0.168866
0.45408
0.373047
0.445129
0.435719
0.439172

Mean value
0.291773

Appendix G
Generating the Normal Distribution with Excel

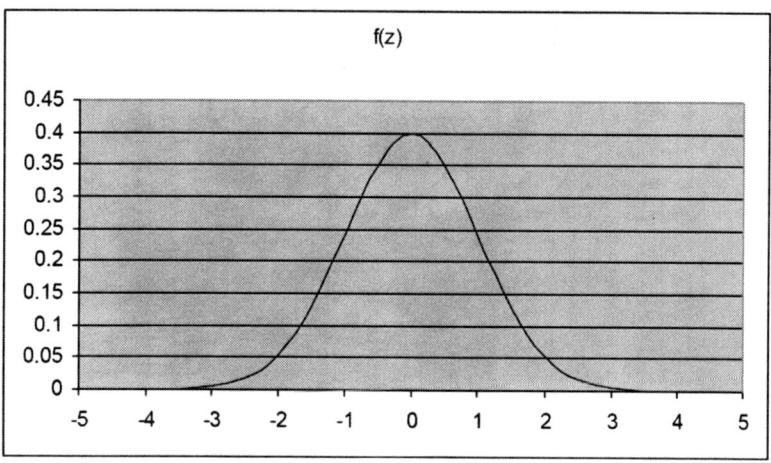

On a blank Excel spreadsheet enter the formula '=((1/SQRT(2*PI())))*EXP(-(B2*B2)/2)' in cell C2 after entering the z values shown on the last page of this appendix in cells B2 through B34. Cells C3 through C34 can then be filled either using auto-fill or manually clicking and dragging the formula in C2 to the remaining cells in the f(z) array.

Now select a convenient cell somewhere near the two data columns just entered. Click the 'Chart Wizard' icon. Then select the 'Scatter chart type' icon.

Next, select the highlighted 'Chart sub-type' icon. Click 'next.' A menu appears asking for the data range. Click and drag the cursor over the two columns of cells originally entered under the labels 'z' and 'f(z)' to highlight them. Click 'next' again, and the desired normal curve appears in a 'Source Data' window.

Click 'next' again, and a 'chart options' window showing 'step 3 of 4' appears. Now you may add titles, etc. if desired. Click 'next' once more and a 'chart location' window appears. Select one of the two 'radio button' options, as desired. Click 'finish' and the customized normal curve appears inside its own frame. Some features may still be altered as preferred.

Appendix G - continued
z and f(z) Values for Generating Normal Distribution

z	f(z)
-4	0.000134
-3.75	0.000353
-3.5	0.000873
-3.25	0.002029
-3	0.004432
-2.75	0.009094
-2.5	0.017528
-2.25	0.03174
-2	0.053991
-1.75	0.086277
-1.5	0.129518
-1.25	0.182649
-1	0.241971
-0.75	0.301137
-0.5	0.352065
-0.25	0.386668
0	0.398942
0.25	0.386668
0.5	0.352065
0.75	0.301137
1	0.241971
1.25	0.182649
1.5	0.129518
1.75	0.086277
2	0.053991
2.25	0.03174
2.5	0.017528
2.75	0.009094
3	0.004432
3.25	0.002029
3.5	0.000873
3.75	0.000353
4	0.000134

Appendix H
The t-Test for Two Samples with Equal Variances

The formula for calculating the *actual value* of t is given below. Note that the value obtained from this process is *not* the probability value α that Excel returns when a significant difference between means is tested. That number is always between 0 and 1, as probabilities must be.

However, the actual t value allows one to determine by how much the difference between sample means must be changed to achieve significance for *any desired probability* value. In that respect, this is a highly important and useful result.

The formula for t when variances are judged equal using the F test is:

$$t = (\bar{x}_1 - \bar{x}_2) \div [(s_p \cdot \sqrt{(1/N_1 + 1/N_2)}]$$

For clarification, the entire expression inside the square brackets divides the numerator, which is the difference between the sample means, $(\bar{x}_1 - \bar{x}_2)$. S_p is termed the *pooled* variance of the two samples being compared.

It is appropriate to use a pooled (read 'combined') variance in this case because it is presumed that the investigator first used the F test to assure the samples have equal variance, or at least variances that do not show a significant difference between them.

In actuality, samples seldom have literally equal variances, but the variances must test equivalent using the F test.

The actual standard deviations, s_1 and s_2 of the samples being compared are each squared and combined in the following formula to derive S_p:

$$S_p^2 = [(N_1 - 1) \cdot s_1^2 + (N_2 - 1) \cdot s_2^2] \div (N_1 + N_2 - 2)$$

Note particularly that after S_p^2 is calculated using this formula the square root of that result, $\sqrt{S_p^2} = S_p$, must be taken and that value then substituted in the first formula for t.

The entire expression $(N_1 + N_2 - 2)$ in the denominator of the formula for S_p^2 is called the degrees of freedom for the t-test. It is denoted generally by df or v.

The degrees of freedom equal the sum of the sample sizes minus 2. For instance, if sample 1 contains 5 values ($N_1 = 5$) and sample 2 contains 3 ($N_2 = 3$), then the degrees of freedom for this example are df = v = (5 + 3 − 2) or 6. Degrees of freedom are critical to obtaining an accurate value for t.

Some texts (Walpole 155), for instance, use the symbol v (Greek lower-case nu) in place of df to represent degrees of freedom. It is best to recognize both symbols.

As a brief example of using the t-value formula, assume that you have two samples with combined df or v = 7. Also assume that you wish to know the value of t that will just make the difference between the sample means significant for the one-tailed t-test at the 0.05 level.

Referring to any table of t values, you will find that for 7 total degrees of freedom and the 0.05 significance level, t must equal 1.895. Suppose further that the two samples have respective means of 71.27 and 70.47 and that their variances do not differ significantly when the F test is run, being $s_1 = 0.72$ and $s_2 = 0.42$.

You can then easily solve the original t-value equation for the difference in the means and obtain:

Appendix H

$$(\bar{x}_1 - \bar{x}_2) = t \cdot [(s_p \cdot \sqrt{(1/N_1 + 1/N_2)}]$$

Since t, s_p, the sample sizes N_1 and N_2 and both means are known, you can determine the exact difference between the sample means necessary to obtain the significance level you wish, within small errors due to rounding.

If $\bar{x}_1$ is the mean (71.27) that must be adjusted, solve the above equation for $\bar{x}_1$ and you find that its final value must be

$$\bar{x}_1 = \bar{x}_2 + P,$$

where P is the entire expression involving t, s_p and the sample sizes, N_1 and N_2 on the right side of the previous equation. Therefore, since $\bar{x}_2$ is 70.47, solving this equation for $\bar{x}_1$ tells you how much to *add or subtract* from 71.27, as the case may be, to arrive at the significance level you wish, within a small rounding error. It is a straightforward application of basic algebra at this point.

Interpret t tables carefully. Be sure to locate the proper t value for the significance level desired and not the value for a different probability level.

For example, the value of t used in this case, 1.895, is for df = 7 and α = 0.05 for one-tailed significance. However, if you wish the corresponding value of t for the same problem using a *two-tailed* significance test, then α/2 = 0.025 (multiply the one-tailed significance level, 0.05, by 1/2 to find the corresponding α level in each of the two-tails). In other words, the total desired alpha level, 0.05, remains the same but is now divided equally between the two tails. The correct t value is 2.365.

The question of finding the appropriate t value is probably the only potentially confusing point, for many, regarding the use of t tests.

The t-Test for a Single Correlation Value, r

A straightforward test exists for determining whether a single value of r calculated between two sets of data differs significantly from zero. The null hypothesis we are testing now is: H_0: $\rho \leq 0$. It reads: rho is less than or equal to zero. In this case our sample correlation coefficient, r, is substituting for the corresponding population correlation coefficient, ρ.

The alternative hypothesis, for this case, is H_1: $\rho > 0$. A one-tailed test is thus used.

The formula is:

$$t = r \cdot \text{SQRT}(N - 2) \div \text{SQRT}(1 - r^2)$$

In the formula, r is the value of a correlation coefficient (either positive or negative) calculated between any two sets of data, N is the sample size and N – 2 is the degrees of freedom, df.

The term 'SQRT' is Excel's acronym for square root. The entire expression 'r · SQRT(N – 2)' must be divided by 'SQRT(1 – r^2)' to determine t.

In Chapter 7 the correlations for 91 years (the sample size) of the yearly running times for the Travers between the *year number* in which it was run and the *foal crop size* for that year were determined.

The respective correlations, r, were -0.82 and -0.64. We wish to determine whether these r values differ significantly from zero. That is, are they significantly greater than zero, where only their absolute magnitude, since the minus sign shows just their inverse direction, need be considered?

When the indicated formula calculations are made, first using r = -0.82 and then using r = -0.64, and with N – 2 = 89, the results are: t_1 = -13.51 and t_2 = -7.86, rounded to two places.

Appendix H

Using Excel's TINV function with probability level 0.10 and degrees of freedom = 89 automatically returns the one-tailed equivalent t value at the 0.05 alpha level (Excel 2003 is designed to give the two-tailed value if 0.05 is entered in TINV). The result, rounded to three decimal places, is t = 1.662.

Since both t_1 and t_2 are definitely larger (the minus signs do not matter for this calculation) than 1.662, both r values significantly differ from zero. Therefore, the Travers correlations are considered meaningful.

Summary on how to Interpret TINV in Excel 2003

If Excel's Help selection is made from the main menu bar and TINV is accessed, the explanation offered is that the probability entered in the topmost of the two fields in the Function Arguments Dialog Box results in the *two-tailed* value of t being returned.

It further explains that, *if the one-tailed t value is desired*, entering *twice* that same probability value returns the *one-tailed* t value for the same 'degrees of freedom,' generally denoted 'df' in text books.

Therefore, if the two-tailed t value for testing a null hypothesis at the $\alpha = 0.05$ level for df = 5 is desired, enter '0.05' in the first field and then enter '5' in the field beneath it, the 'Deg_freedom' field.

The value t = 2.571 is returned. Checking this against the table values in Appendix E, shows that 2.571 is the t value opposite 5 degrees of freedom and under the 0.025 probability-value column.

This is consistent because if one is testing a difference between means using a total two-tailed probability of 0.05, then one-half that probability value, 0.025, will be in each of the two tails. This actually means that t = ± 2.571 is the appropriate interpretation so that the test can cover both the right and left tails of the distribution..

If a one-tailed t value is desired at the same overall probability level, 0.05, then enter *twice that value*, 0.10 in the uppermost, or 'Probability' field, of TINV.

Using the same df, 5, results in t = 2.015 being returned. This is the table value from Appendix E opposite 5 degrees of freedom and beneath the 0.05 column. This is also consistent since you are asking for a one-tailed t value when df = 5 and α = 0.05.

Although everything is consistent in these results, the TINV function still seems awkward to use without compromising accuracy of interpretation.

Appendix I
Justification for the 50/50 Division of Track Variant

In chapter 6 the section entitled "The Sequence of Time adjustments," stated that 50 percent of Track Variant may be considered due to surface effects per se and the remaining 50 percent due to quality variations in the runners.

The following is now offered as justification:

Four races considered the most prestigious in America were analyzed for the effects of foal crop and track variant on running time.

Sixty consecutive years, 1917 through 1977 of Travers data were used to determine the correlation and COD between running time and foal crop. These are the years which include the foal crops of Man o' War (1917) and Secretariat (1970).

The correlation between those variables was -0.5833 with a COD of 0.3403. These numbers mean that as foal crop increases running time decreases (indicated by the negative sign) and that thirty-four percent of the changes in run time are explained by the changes in foal crop size.

It also implies that about 0.66 of running time (subtracting 0.34 from 1.00) is attributable to factors *other than* quality of the runners, since we are assuming that larger foal crop is proportional to increase in number of quality runners. This sixty-six percent may be attributed to track and related conditions.

Second, the Kentucky Derby, Preakness and Belmont were analyzed with respect to the Triple Crown winners, excluding Sir Barton and Gallant Fox. The latter two colts won "in retrospect" when Omaha was the first horse recognized for winning all three races at their current distances in 1935.

It was felt that trying to compensate for the variation in distances run by Sir Barton would introduce data contamination. This left ten actual winners of the Triple Crown beginning with Gallant Fox in 1930 and ending with Affirmed in 1978.

The resultant correlations between running time and track variant for the thirty collective Triple Crown races of these champions were: Kentucky Derby: 0.6772; Preakness: 0.7137; Belmont: 0.8024.

The corresponding CODs for these correlations are: 0.4586; 0.5094 and 0.6438.

The average COD for the Triple Crown races plus the 0.66 Travers value is 0.5680.

Therefore, it is certainly justifiable to assume that at least 0.50 or 50% of Track Variant is attributable to track surface (track variant effect on) and the remainder to quality of foals in the respective crops, based on these four prestigious races.

Since these races generally featured the highest quality in Thoroughbred runners, questions about the quality factor are essentially eliminated from consideration. Thus, this result is close to a "true," but never exactly discernable, value for track effect.

Appendix J
Time versus Impost Trends: Six Standard Distances

Times, in seconds, are plotted along the vertical axes; imposts, in pounds, are plotted along the horizontal axes. Each separate trend equation, R^2 value and sample size is shown beneath its corresponding chart.

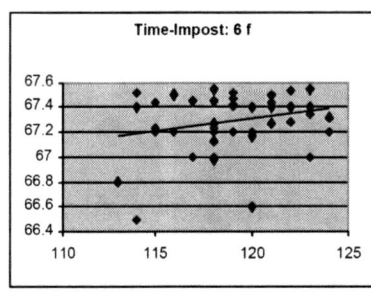

$y = 0.0199x + 64.922; R^2 = 0.065; N = 48$

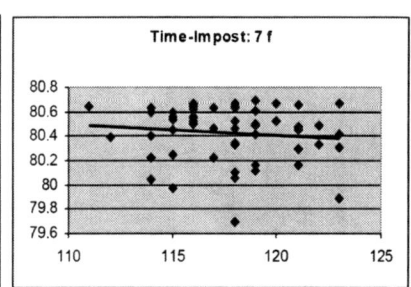

$y = -0.0087x + 81.45; R^2 = 0.012; N = 54$

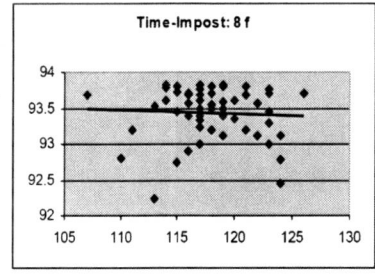

$y = -0.0053x + 94.053; R^2 = 0.003; N = 54$

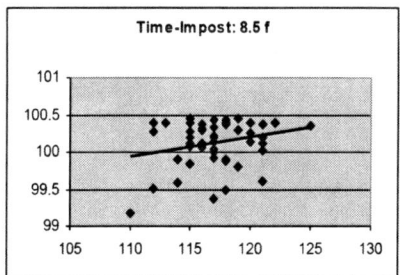

$y = 0.0262x + 97.049; R^2 = 0.060; N = 50$

Appendix J continued

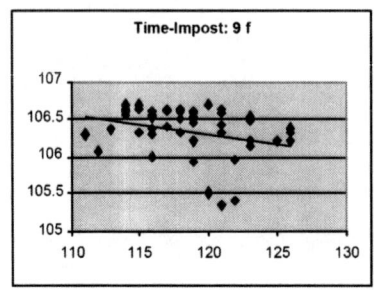 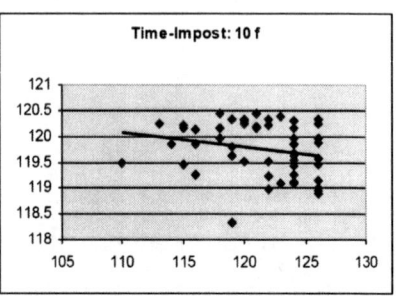

y = -0.025x + 109.303; R^2 = 0.084; N = 52   y = -0.0271x + 123.056; R^2 = 0.048; N = 54

Appendix K
Calculating Confidence Intervals

When a predicted value, $\hat{y}_0$, is found using a linear regression equation, it represents only half the necessary information. To be useful, the value $\hat{y}_0$ must be accompanied by a confidence interval, generally abbreviated CI.

For reference, if needed, $\hat{y}_0$ is generally called 'y-sub-zero hat' by statisticians. This is undoubtedly one of those rare instances of attempted humor, referring to the circumflex above y as a hat, by statisticians ever recorded!

The standard values of confidence intervals include 90, 95 and 99 percent. If, for example, an average value for some sample has been calculated as 40.5 and a 90-percent confidence interval for that value is ± 3.75, it means that, if one takes *repeated random samples* of identical size, n, *from a given population*, then the average value for 90 percent of those samples will be somewhere between 36.75 and 44.25 (40.5 ± 3.75).

By analogy, the 95- and 99-percent confidence intervals work identically. The difference is that the higher percent confidence one has in the value range for a certain statistic, the wider that range is. There is no such thing in statistics as a free lunch.

The following is a condensed presentation of how to calculate a confidence interval for a given predicted value obtained using linear regression.

The following example data represent six actual **distances**, in furlongs, and **run times**, in seconds, which are the current world records on dirt for those distances. All data are from the current 2012 edition of the ARM. (911) The output table from the LINEST application of Excel is shown immediately adjacent and to the right of the sets of independent and dependent variables, distance and run time, respectively.

Distance	Run Times on Dirt	LINEST Output Table	
6f	66.49	12.81429	-10.36714
7f	79.40	0.028856	0.236268
8f	92.20	0.99998	0.092195
8.5f	98.40	197208.1	4
9f	105.00	1676.269	0.034
10f	117.80		

The general formula for calculating a confidence interval was given in the text as $\hat{y}_0 \pm A$, where A is actually one-half the entire CI width. The *complete expression* for A is:

$$A = \pm [t_{\alpha/2} \cdot s \cdot \text{SQRT} ((1/n + (x_0 - \bar{x})^2 \div \Sigma_6 (x_i - \bar{x})^2))].$$

In the formula, $t_{\alpha/2}$ is the value, from a table of t values, *for the regression degrees of freedom, n – 2, and the desired confidence interval*; n – 2 is given in row 4 on the right side of the LINEST output table shown above; s is the *standard error of estimate* in the same output table and directly above the n – 2 value; 1/n is the reciprocal of the sample size, n. It is 1/6 for this example; x_0 is the value of the *independent variable, x, for which $\hat{y}_0$ is to be determined* from the regression equation; $\bar{x}$ is the mean of the six x variables, and x_i denotes *each* of the six x-variable values taken individually and successively.

The highlighted area in the equation for A stresses that the *entire expression* is contained in the same parenthetical group and *then* the square root (SQRT) on that expression is taken. Within the highlight, however, 1/n is *not included* in the indicated division, '÷'.

Appendix K

The Σ_6 means that there are *six separate values*, in this example, of $(x_i - \bar{x})$ to be squared *and then* summed *before* the division is completed.

For easy reference the specific values just explained are, for the present example: $t_{\alpha/2} = 4.604$; df $= (n - 2) = 4$; s $= 0.092$; $1/n = 0.167$; $x_0 = 8.0$; $\bar{x} = 8.083$; $(x_0 - \bar{x})^2 = 0.007$; $\Sigma_6 (x_i - \bar{x})^2 = 10.208$.

Combining all these values in the above equation for a 99% confidence interval, A, gives: A $= \pm 0.174$. All values in this example, other than for x_0, have been rounded to three decimal places. Dividing A by 2.58, the z-value for a 99% CI, then gives the standard deviation associated with this CI, that is, $0.174 \div 2.58 = 0.067$.

When the value $x_0 = 8.0$ is substituted in the regression equation $12.814 x_0 - 10.367$ (from the top highlighted line of the LINEST output table), the value $\hat{y}_0 = 92.147$ is obtained. This is extremely close, as is expected, to the actual value of 92.2 s (which happens to be Dr. Fager's world-record time for 8f) when x is 8f. The *error of estimate*, also called *residual*, is 0.053. This value is well within s $= 0.092$, the *standard error of estimate* given by the output table.

This accuracy is also expected, since the COD for the regression equation is 0.99998, as shown in row 3, column 1 of the output table. It implies that changes in running time, even at world-record level, are predicted with over 99.99 percent accuracy by changes in distance run.

The above example hopefully makes the admittedly time-consuming process of calculating confidence intervals more palatable.

Man o' War & Secretariat

Appendix L
Adjusting Linear Trends to World Record Trend

Figure L.1 displays portions of three trend lines. The upper line is taken from Man o' War's juvenile data for five, five and one-half and six furlongs. The middle line is from Secretariat's six-furlong data, and the lower line represents the 2011 North American dirt track record data for six furlongs published in the American Racing Manual.

There is a relatively simple method of adjusting the *entire set* of Man o' War's run times to world-record times for the same distances. The same adjustment is used for Secretariat.

This type adjustment should be at least as accurate as the race-by-race adjustments performed originally. It also serves to corroborate the previous results because it preserves the *same relative pattern and separation of data points* and assures that all such points are adjusted equally with respect to the 2011 world records.

The adjustments are made in this case by lowering Man o' War's and Secretariat's trend lines until the *areas* beneath them *exactly equal* the area beneath the world-record trend line *between the same end points*.

The areas beneath all trend lines are trapezoids with overall boundary ABCD, shown in Figures L1 and L2. Line AD is the zero-reference or x-axis for all numbers. It also represents the *heights* of all trapezoids discussed. The figures are not drawn to scale.

This area convention is also used, for example, in basic physics regarding work. Work is physically defined as the *product* of a *force* (in pounds or Newtons) multiplied by the *displacement* (distance) it moves an object (in feet or meters) *along the same direction* the force acts. The *area under the force-displacement graph* (in pound-feet or Newton-meters) is then used as a measure of the total work done – precisely analogous to the use here of height in furlongs multiplied by times in seconds to define an area of interest. (Sears 118)

The area beneath the six-furlong world-record trend line is calculated using the standard formula for the area of a trapezoid, 0.5 x height x (sum of bases). Numerically this is 0.5 · (1.00) · (61.03 + 73.00) = 67.02. The actual area units need not be expressed if consistency between all applications is maintained.

The area beneath Man o' War's regression is also trapezoidal and is found by using the same formula: 0.5 x height x (sum of the bases). In this case the height is 1.00, the same as for the world-record trend line.

Thus, the area beneath Man o' War's trend line that must be adjusted to that beneath the world-record trend line is: 05 · (1.00) · (66.22 + 78.17) = 72.20.

The difference between these two areas is: 72.20 – 67.02 = 5.18. Therefore, Man o' War's trend line must be lowered just enough so that his 72.20 area units are reduced by 5.18 area units.

The required adjustment may at first seem complicated, but one only need realize that, since the height (horizontal baseline) of the trapezoid to be adjusted is 1.00, then lowering the entire trend line by 5.18 units reduces the area beneath it by the required amount because 1.00 x 5.18 = 5.18, and this lowering has effectively subtracted 5.18 area units from the original total area..

Appendix L

The same procedure is done for Secretariat's linear trend line representing his six-furlong races. The area under Secretariat's two-year-old linear regression also extends from 5.5 f to 6.5 f along the x-axis, as it did for Man o' War's and the world-record line.

Therefore, Secretariat's six-furlong area relative to the world-record line is: $0.5 \cdot 1.00 \cdot (64.02 + 76.95) = 70.49$. When Secretariat's area is reduced by the difference between the world-record area and his the result is: $70.49 - 67.02 = 3.47$.

The *overall effect* is the same as calculating the direct difference between Man o' War's area reduction and Secretariat's. The result is the amount that must be subtracted from each of Man o' War's six-furlong times to make them compatible with Secretariat's with both data sets referenced to the world-record.

The final result is: $5.18 - 3.47 = 1.71$ s (including time units).

It is interesting that this result essentially confirms that it is reasonable to use the trapezoidal approximation. The reason is that a direct subtraction of Man o' War's and Secretariat's *average times* for six furlongs gave: $72.21 - 70.47 = 1.74$ s. The difference between the two methods is only 0.03 s.

Summary results only are now given for the nine-furlong calculations since they exactly parallel those for six furlongs. The reader is urged to complete this example on his own for practice and to assure understanding.

The area under the nine-furlong world-record line is given by: $0.5 \cdot 1.0 \cdot (96.40 + 112.40) = 104.40$; the area under Secretariat's trend line is: $0.5 \cdot 1.0 \cdot (101.13 + 113.97) = 107.55$. The required adjustment to match the world-record area is: $107.55 - 104.40 = 3.15$.

Therefore, this is the amount that Secretariat's trend line must be adjusted downward to match the world-record area for nine furlongs.

The area beneath Man o' War's 9f regression line, and including the same baseline limits, is: 0.5 · 1.00 · (104.37 + 116.79) = 110.58. Direct subtraction of the world-record area from his area gives: 110.58 − 104.40 = 6.18.

Therefore, the *relative amount* by which Man o' War's times must be adjusted to Secretariat's is: 6.18 s − 3.15 s = 3.03 s. This difference represents a little more than one second difference from Farley's estimate for one mile. It is actually less than a one-second disagreement since this calculation is for an extra one-eighth of a mile.

Refer to the following figures in which all numerical values are understood to represent seconds.

Figure L.1

Six-Furlong Trend Lines: Trapezoid Adjustment

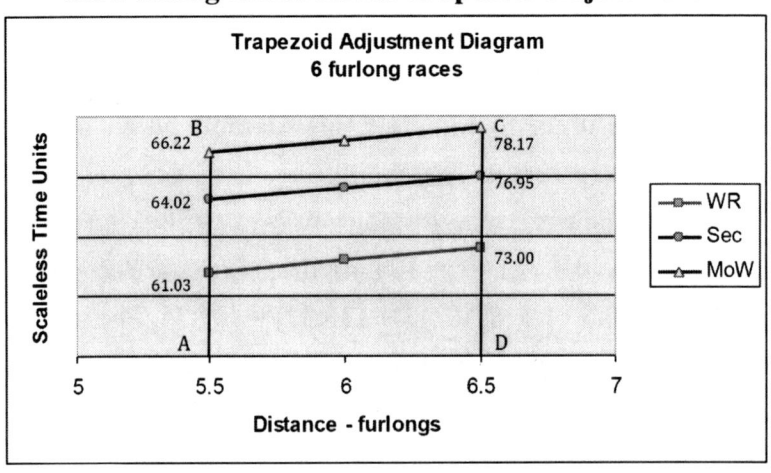

Appendix L

Figure L.2
Nine-Furlong Trend Lines: Trapezoid Adjustment

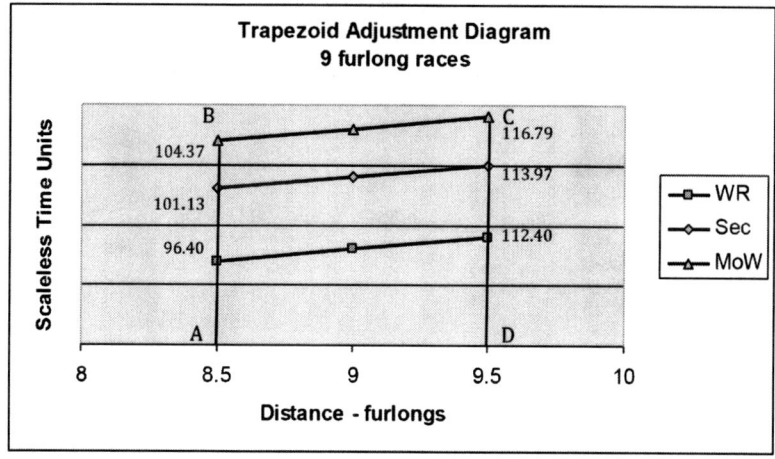

Index

Acceleration, 8
    physical units for, 8
Adjustments (time), 200
    juvenile year, 129
    sophomore year, 200
Affirmed and Triple Crown, 230
Alpha error, 32
Alternate hypothesis, 35
Angle Light and Wood Memorial, 231
August Belmont, Jr., 55
Average (statistical), 15
Average speed, 7
Baffert, Bob, 93
Belmont and Secretariat, times per furlong, 233
Beta error, 35
Bold Ruler, 57
Bold Ruler and Secretariat, 57
Byerley Turk, 56
Canadian International Stakes, 84
Carbon dioxide dissipation and inheritance, xxvii
Chenery, Christopher, and The Meadow, 81
Citation and the Triple Crown, 92
Coding juvenile correlation sequences, 117
Coefficient of determination (meaning), 28
Combinations, calculation of, xxvii
Comparisons, juvenile for Man o' War and Secretariat, 117
Confidence interval general formula, 155
Coordinate notation, 26
Correlation, xix, 19-20
    Pearson coefficient, 24-25
    of dice rolls, 20
    between Man o' War and Secretariat as juveniles, 120-121
Cougar II, 100
Curwen's Bay Barb, 56
Damascus and American Derby record, 98

Darley Arabian, 56, 227
Data for sophomore year, 187
    sources of, xx
DeMoivre, Abraham, and normal distribution, 16
Devonshire 'Flying' Childers and world record comparison, 40, 45
Distance as used in racing, 4
Doswell, Virginia, 81
Dr. Fager and one-mile world record, 45
    Work-Energy Theorem, 45
Eddie Maple, 84
    Secretariat in the Canadian International, 84
Eight, as number for genetics, xxvi
Era difference, 39, 43
    Flying Childers, 40-41
    effects of and z-scores, 173-174
Evolution and time span of, xxv
Feliciano, Paul, 84
Flying Childers, 40, 144
    relative to current world record, 41
    linear regression for, 42
    relative to Work-Energy Theorem, 45
    compared with Dr. Fager, 44
Foal crop, 49
    running ability, xxvi
    Thoroughbred talent, 47, 50
Fokker Dr. I tri-plane and era comparisons, 135
Furlong and equivalent lengths, 4
Gaffney, Jimmy, and Secretariat's workout problem, 90
Gauss, Carl Friedrich, and the normal distribution, 16
Gene pool of Thoroughbred, xxv
General Assembly, 103
    Travers record of, 103
Gilman, Dr. Manuel, veterinarian, 91
Godolphin Arabian, 56
Greatness, and inherent running ability, xv, 236
Hasty Matelda and Something Royal, 83
Heart size inheritance and the X-factor, xxvii

Index

Hellcat, Navy F6F and era comparisons, 135
Hoodwink and the Lawrence Realization, 72
Hypotheses, statistical, 32
    acceptance of for statistical tests, 35
    rejection of for statistical tests, 35
    testing of and random sampling, 29
    relative to the t-test, 29, 31
Impost adjustment and the juvenile year, 151-152
Inheritance and natural odds, xxv
Insert function button, Excel 24
Juvenile data, 108
    time adjustments and related raw data, 129-133
Kinetic energy, 45
Lactic acid dissipation and inheritance, xxvii
Lady's Secret, 103
Lang, Chick, general manager of Pimlico Racetrack, 102
Laurin, Lucien, vii, 91
linear regression, 25
    abscissa, 21
    dependent variable, 25
    independent variable, 25
    intercept, 28
    related to least squares, 27
    ordinate, 21
    standard error of estimate, 28
    time prediction for Belmont call points, 233-234
    trapezoidal time adjustment for juvenile year, 159
    meaning and interpretation of, 25-26
    between-race predictions, 230
    within-race predictions, 232
Lukas, D. Wayne, 104
Lumenaut as Excel add-on statistical package, xx
Lung capacity and inheritance, xxvii
Man o' War, general biography, 55
    Kentucky Derby, 69
    rest period between six-furlongs races, 112-113
    related to Secretariat by ancestry, 56-57

specific ancestry, 55-56
August Belmont, Jr. as breeder of, 55
Belmont Stakes, 233
    Secretariat, 233-235
    Man o' War, 236-239
boundaries for juvenile time adjustments, 129-133
call point 1 times for six furlongs, 112
Dwyer Stakes, 75
related to Eclipse, 56
field size for six-furlong races, 111
finish margin for six-furlong races, 110
imposts for six-furlong races, 111
interment, 250
John P. Greer and, 76
juvenile running style of, 63
juvenile temperament of, 58
Kelso and, 71
Lawrence Realization, 71
maiden race, 58-59
overall gain for six-furlong races, 112
post position for six-furlong races, 111
retirement of, 78
Sanford Stakes, 60
Secretariat and sophomore comparison with, 187
seventy-percent win ratio for nine furlongs of, 210-212
    calculation of, 224-225
Sir Barton match, 77-78
sophomore anomaly, first, 69
speed rating for six-furlong races, 111-112
time correlations for six-furlong races, 110
times per furlong of Lawrence Realization, 74
track condition for six-furlong races, 111
Track Variant for six-furlong races, 52 113
Will Harbut, 78-79
Will Harbut and Saturday Evening Post, 80
Belmont comparison with Secretariat, 236-238
juvenile time correlations, 118
foaling, 55

Index

Martin, Frank, and Belmont Stakes, 95
    Sham, 95
Mass vs. weight, 6
Mean (statistical), 15
    sample, 15
Mendel, George, and genetics foundation, xxvi
Metabolic factors and treadmill testing, xxvii
Miss Disco and Bold Ruler, 57
Momentum, 10
Monarchos and Kentucky Derby, 93
    Preakness Stakes, 93
Muscular strength and inheritance, xxvii
My Gallant and Arlington Invitational, 98
Nack, William, 90
Nine-furlong races, 189-194
    adjusting times of, 198
    setting time boundaries for, 207
Nominal race distance defined, 4
Normal curve and areas beneath, 16-18
Normal distribution, 16
    generating with Excel, 17
Normalcy test, Shapiro-Wilk, xx, 19, 108
North American Dirt records graph, 41
    graph of, 42
Null hypothesis, 32
Ockham's razor, principle of, xxxii
Our Native and Arlington Invitational, 97-98
Oxygen transport efficiency and inheritance, xxvii
Oxygen uptake efficiency and inheritance, xxvii
Peers and quality factor related to Man o' War, 182
Phipps, Ogden, and Secretariat coin toss, 82
Point Given, 93, 94
Population, definition, 11
Preakness Stakes and Maryland Racing Commission, 218
    Secretariat's record time and, 218
Prediction using random simulations, xxv
    six-furlong time adjustments and, 136

Prove Out, winner of Woodward Stakes, 49
Random factors, influence of, xvii
Range of sample, 15
Ratio scale, 11-12
Reid, Susan, 99
Risen Star, 103
    Triple Crown and, 103
Riva Ridge, 101, 235
Sample definition, 12
    large sample, 13
Sanford Stakes and Man o' War, 60-66, 69, 124
Sarah Jeffords and Lawrence Realization, 72
Scalar quantity, 7
Seconds abbreviation, 5
    unit of time measurement, 5-6
Secretariat and rest period between six-furlong races, 98
    "zinger" workout, 100
    abscess and Wood Memorial, 90-92
    ancestry, 83-84
    anomalies, career, 90
    Arlington Invitational, 97-98
    Bay Shore, 87-88
    Belmont Stakes, 233
    Belmont Stakes per-furlong world-record comparison, 235
    Call Point 1 times for six-furlong races, 116
    Canadian International Stakes, 101-102
    Feliciano, Paul, jockey, 84
    field size, six-furlong races, 116
    final six career races, 97
    finish margin for six-furlong races, 116
    foaling of, 82
    General Assembly, 103
    Gotham Stakes, 88-90
    impost for six-furlong races, 115-116
    interment, 105, 250
    Kentucky Derby, 93
    Lady's Secret, 103

Index

Laminitis, 105
maiden race, 84
Man o' War six-furlong comparisons, 108-114
Man o' War Stakes, 100
Marlboro Invitational Stakes, 99
overall gain six furlongs, 117
post position, six-furlong races, 116
progression of races, 88
Preakness Stakes, 94
    time made official, 93
running style, 85-86
Shapiro-Wilk normalcy tests, 108
sophomore season, 86-88
speed rating for six-furlong races, 115
time correlations for six-furlong races, 114-115
times for consecutive Belmont furlong splits, 96
track condition for six-furlong races, 117
track variant for six-furlong races. 115
Triple Crown, 92
Whitney Stakes, 99
Wood Memorial Stakes, 91
Woodward Stakes, 100
sire of champions, 103-104
correlations for juvenile year, 109, 114-117
    with juvenile times, 115
foaling, 82
juvenile year data, 108
Set, meaning of, 15
    symbol for, 15
Sham and Belmont Stakes, 95-96
    second fastest Kentucky Derby, 93
    second fastest Preakness Stakes, 94-95
Shapiro-Wilk test, xx, 19, 108
    normalcy check for nine-furlong races, 197
Significance level, α, 32
Significant differences for nine-furlong races, 194-196
Simply Majestic and nine-furlong world record, xxxi

Simulations, fluctuations of Track Variant, 51-52
    how to perform using Excel, 171-173
    juvenile year, 169
    random factors, six-furlong races, xxv
Six-furlong races, xxiii
Skeletal soundness and inheritance, xxvi
Slew, Seattle, and Triple Crown, 230
Slope, meaning of, 22
Something Royal, 83
Sophomore races, xxiii
    adjustments and simulation results, 208-210
    data and raw-form comparisons, 202
    race parameters and time correlation, 188-193
    time adjustments sequence, 200-202
    data adjustments for comparability, 198-199
    procedure for comparisons, xxiii
Speed Rating definition, 50
Stakes winners, odds for, xxviii
Standard deviation, 15
    sample, 15-16
Standard normal distribution, 17
Statistics and ethical applications, 247-248
    descriptive, 14
    inferential, 14
    limitations of, xv
Stockings (white) and running ability, 225-227
Storm Cat, 103
Sweat, Eddie, 90
Terlingua, 103-104
The Meadow, 81
Thoroughbred racing terms, 4-10
Thoroughbred talent and foal crop, 49-50
Thoroughbred uniform running ability, 140-142
    improvements in, 47
Three-sigma limits definition, 18
Time as physical quantity, 5-6
    adjustments and juvenile data, 129-140

Index

  adjustments for juvenile times, Man o' War and Secretariat, 129-134
  sophomore boundary favoring Man o' War, 210-215
Track speed, 239
Track Variant and 50-50 split, 146-147
Track Variant definition problems, 50-52
  estimating running ability and, 50
Trapezoidal adjustment method for sophomore year, 217
Trip, definition and distance, 5
Triple Crown first established, 229
t-Test, xix
  general formula for calculating, 32
  random sampling results and, 30
  related F test for equal variance, 33-34
  significance level using, 32
  testing sample differences with, 29
  interpretation of, 35, 36
  results of and reference baseline for sophomore year, 202-207
  Student's, 29
Turcotte, Ron, 84
  abscess problem and Wood Memorial, 92
  quote about Secretariat toying with field, 100
Tweedy, Penny, and coin toss, 82-83
Twin Sparks and six-furlong world record, xxx
Type I error, 35
Type II error, 35
Variance of sample, 15
Vectors, definition, 7
Velocity, definition, 9
Viscoride track energy absorption, 241
Whittaker, Edwin, and Wood Memorial, 91
z-Scores, and juvenile comparisons, 175-180
  relation of sophomore data to nine-furlong world record using, 220-222
  world-record comparisons using, 183-185
  derived from specific race results, 175-181

CPSIA information can be obtained at www.ICGtesting.com
Printed in the USA
BVOW03s0536090414

350157BV00010B/402/P